Taken By Storm

The Troubled Science, Policy and Politics of Global Warming

Christopher Essex and
Ross McKitrick

KEY PORTER BOOKS

National Library of Canada Cataloguing in Publication
Essex, Christopher
        Taken by storm : the troubled science, policy and politics of global warming / Christopher Essex and Ross McKitrick.

Includes bibliographical references and index.
ISBN 1-55263-212-1

1. Global warming. 2. Global warming—Government policy.
I. McKitrick, Ross, 1965– II. Title.

QC981.8.G56E73 2002   363.738'74   C2002-905025-1

THE CANADA COUNCIL | LE CONSEIL DES ARTS
FOR THE ARTS | DU CANADA
SINCE 1957 | DEPUIS 1957

ONTARIO ARTS COUNCIL
CONSEIL DES ARTS DE L'ONTARIO

The publisher gratefully acknowledges the support of the Canada Council for the Arts and the Ontario Arts Council for its publishing program.

We acknowledge the financial support of the Government of Canada through the Book Publishing Industry Development Program (BPIDP) for our publishing activities.

Key Porter Books Limited
70 The Esplanade
Toronto, Ontario
Canada M5E 1R2

www.keyporter.com

Cover design: Jack Steiner
Electronic formatting: Jean Peters
Author photos: Peter Poole (C. Essex); Michael Levy (R. McKitrick).

Credits: Figure 2.2, photo copyright (2002) by John L. Daly, used by permission; Figure 6.1, copyright the Intergovernmental Panel on Climate Change, used by permission; all of the cartoons were drawn by Christopher Essex.

Printed and bound in Canada

02 03 04 05 06 07 6 5 4 3 2 1

To our grandfathers:
Ants Rehtlane (CE)
Malcolm Omond and Bruce McKitrick (RM)

Although we never met them, we have always known them.

Contents

Acknowledgements   7

Preface   9

1   A Voice from the Whirlwind   14
2   The Convection of Certainty   21
3   Climate Theory Versus Models and Metaphors   61
4   T-Rex Devours the Planet   119
5   T-Rex Plays Hockey   154
6   The Unusual Suspects   175
7   Uncertainty and Nescience   207
8   Ceiling-Fan Gases and the Global Blowing Crisis   233
9   Kyoto, Son of Doctrine   268
10   After Doctrine: Making Policy Amid
      Uncertainty and Nescience   289

Glossary   312
Index   315

Acknowledgements

We are grateful to the Donner Canadian Foundation, and Patrick Luciani in particular, for financial support of this project. We would also like to thank Michael Walker of the Fraser Institute in Vancouver, B.C., for his generous gifts of time, advice and practical help throughout the publishing process. Natalie Clayton and James Wishart provided able research assistance. Anastasios Tsonis kindly read through the finished manuscript on short notice. Finally, we thank our families for their patience during the long intervals where "the book" seemed to crowd out a lot of more important things.

Preface

"In the realm of the seekers after truth there is no human authority. Whoever attempts to play the magistrate there founders on the laughter of the Gods."
—Albert Einstein

You've heard it, we've heard it: Global warming is the greatest threat facing humanity. One hundred Nobel laureates recently signed a statement saying so. A UN panel of scientists says so. Our governments say so. Can all these people be wrong?

Of course they can. Whether the earth is warming or not is a scientific question, not a political one. The scientists who have spoken out about global warming understand this. They know that they can be wrong, and they know that what they believe is not necessarily the same as what is true, no matter how passionate their feelings may be.

If you believe that Nobel laureates can never be wrong, or that UN panels never make mistakes, then be advised that science doesn't work that way. Our first task in this book is to convince you that those Nobel laureates are probably wrong about global warming and that most of what you have been hearing about it is wrong. That will actually be the easy part. The claim that there is a global warming crisis threatening to bring chaos and destruction upon the world is so feeble you were probably feeling somewhat skeptical anyway. You were right—and we'll show you why.

Our second task is to try and explain why so many people, including many smart, famous and high-profile individuals, have gotten so confused on this topic. This will be somewhat more difficult, wrapped up as it is in the complex dynamics at the interface of politics and science, while submerged in a soup of pseudo-science and fear. We believe there has been no conspiracy to dupe the public and the people in question have nothing but the best of intentions. It appears instead that a lot of well-meaning people got locked unwittingly into a game that requires them to speak as if they are absolutely certain about matters upon which certainty is inherently impossible. Many prominent players have staked their reputations on positions that cannot be supported by science or sound policy analysis. Consequently a debate in a free and open marketplace of ideas can't happen. Instead, what we get looks more like a fortress, heavily defended by an arsenal of authoritarian pronouncements designed to intimidate outsiders into staying away. Rhetorical weapons like the statement of the 100 Nobel laureates, and others we have seen like it in the past year, would never be used in a real scientific debate. That big players in this issue feel a need for this sort of cannonade just shows how far we have departed from sensible intellectual practice on what remains, even today, an open scientific question.

This leads us to our third and most difficult task. We will try to look beyond the global warming issue itself and ask how we should make public policy in cases where the underlying science is uncertain. What we end up suggesting will not look anything like the process used around global warming. That will serve as an example of what *not* to do. We will argue for a new approach in which non-scientists stop looking for shortcuts around the hard work of learning the science, and high-ranking scientists stop resorting to authoritarian grandstanding as an easy substitute for the slow work of research, debate and persuasion.

Global warming ceased to be a subject of scientific debate years ago. Watch how critics jump straight to an examination of motives or credentials rather than the substance of an author's argument whenever books like this one are published. The argument, it seems, is that what you say, whether it is true or not, matters less than why you say it or who you are. Scrutinizing the messenger rather than the message may be an effective political ploy, but what we are engaged in is not, at its heart, a political question. Our society has already done substantial harm to itself by not grasping this crucial point.

For those who will find this book threatening, and instinctively grasp for this sort of defence, you're in luck. While we have neither sought nor received any industry funding for writing this book, the fact is our "green" credentials are pretty pathetic. We like civilization and all its artificial material comforts. Trees and lakes are nice to have around, but so are paved roads and warm buildings. We find recycling and the whole blue box thing a bit of a sacred cow. We generally prefer driving to walking, especially when transporting kids, except for very short distances. We don't care if people drive subcompacts or SUVs—that's their business. Live and let drive. We eat meat, use pesticides on our lawns and avoid using public transit if possible. We think people who pay extra to buy "organic" vegetables are basically suckers. One of us even smoked *cigarettes* (for a while, anyway). We have no idea when Earth Day is, nor do we care, as long as the malls stay open.

So there. If you are worried that this book may unravel some of your cherished beliefs about global warming, just gently close the cover and put it back on the shelf. Tell yourself these two horrible people are so out of touch with nature they couldn't possibly be trusted.

For those of you who choose to read on, you still might wonder why we wrote this book. Not that it matters in deciding whether we're right, but one is just naturally curious.

We wrote this book because we got tired of seeing irrational fears about global warming cause nations and their leaders to rush around in a panic about a crisis that in all probability does not exist and enact obscenely expensive policies that would not fix anything even if it did. All the money spent on giant conferences to sign meaningless treaties, and to implement policies that arise from these treaties, could have done so much more good if only people had stopped to think about priorities.

Think about basic things. We live in a world in which people in poor countries die from drinking contaminated surface water. We can't find the money to drill wells for them, yet we pour billions of dollars each year into global-warming-related projects that benefit no one. We live in a world in which people suffer lung diseases from breathing air contaminated by indoor dung fires—a common method of heating and cooking in Third World world villages—yet we waste our money worrying about a harmless gas in the lower troposphere. One of us lives in a city in which a fifth of the residents have no family doctor because there isn't enough money to recruit new general practitioners, yet our government has budgeted over a billion dollars to "fight" global warming on the absurd belief that such spending will prevent "wide-ranging and mostly harmful effects on human health" due to climate change.

We wrote this book because we got tired of seeing science twisted into a prop for political ideology. We have seen the work of intelligent and skilled scientists get attacked as "marginal" and "skeptical" and thereby get forced out of mainstream academic publications, for no reason other than that it raises questions about global warming. We have also seen extremely weak studies sounding the global warming alarm get a red-carpet welcome into top journals.

We wrote this book because we got tired of opening the newspaper or turning on the TV news and seeing a river of idiotic, alarmist nonsense rushing out at the public, and of knowing that there is no way to challenge these reports within the media themselves. We have learned from experience that letters to the editor, and even carefully documented factual challenges to individual journalists, are simply ignored.

We wrote a book because nothing else seemed to work. We got tired of making technical presentations to politicians and bureaucrats who claim they want to get informed about the issue, only to have them resort to phony evasions like the "precautionary principle" to rationalize dismissing evidence and arguments they don't like or don't understand.

And, finally, we wrote this book because we got tired of waiting for someone else to do it. Frankly, good scientists and economists naturally stay away from this sort of project, for understandable reasons. Global warming is a topic that sprawls in a thousand technical directions. There is no such thing as an "expert" on global warming, because no one can master all the relevant subjects. On the subject of climate change everyone is an amateur on many if not most of the relevant topics, including ourselves. Since academics do not like writing on subjects outside their areas of expertise, this has been enough to convince many of our colleagues to quit the field altogether, hoping someone else will come along to clean up the mess.

One hundred Nobel laureates: That's pretty impressive. However, we were not told how many Nobel laureates declined signing the letter. It is very doubtful that all of the big prize winners (and there are quite a few) would agree to sign, as there are many divergent opinions among any such group. Still, it seems almost a sacrilege to take a position against so many of them, especially on a motherhood issue like global warming.

But this is a foolish notion. There is nothing sacred here. In terms of global warming, aside from the other issues they discuss, it is clear that they were motivated by moral concerns. They got the science wrong: "Of these poor and disenfranchised, the majority live a marginal existence in equatorial climates. Global warming, not of their making but originating with the wealthy few, will affect their fragile ecologies most." Yet it's the polar and not the equatorial regions that are most affected in the forecast. Oops!

We can all easily become distracted from the truth of a thing by pomp and majesty. Here is a true story to illustrate.

A few years ago, a Nobel laureate made some gratuitous political remarks that were highly offensive to people at his table at a dinner party. Perhaps the laureate had had too much to drink, or maybe he was just spoiling for an argument. A number of guests were appalled. But the scientists gathered there were so intimidated by the fact that it was a Nobel laureate speaking that they froze like deer in the headlights of a car. A young woman—she was barely 20—at the table, a guest of one of the other delegates, boldly spoke up and sharply rebuked the Nobel laureate. By the time she was done, she had shamed them all, reminded as they were by her courage that people with grand honours are no more than human after all.

Her courage leads the way. We are all just humans together whenever we try to deal with nature. We all face the laughter of the gods when we try to pretend otherwise. That is the spirit in which we write. We make no promises that this will be easy to read, nor do we promise to cover every topic or do justice to every point of view. This is our view. But you will not find it dull. Get ready for a ride into the thunderstorm.

1 A Voice from the Whirlwind

It is a dark and stormy afternoon. The winds howl and the dog cowers while thunder rolls and lightning flashes. A tornado warning is in effect, which means someone has seen one. This wondrous noise of sparkling, booming rain and wind is no man-made thing. Its footprint is about 20 kilometres across, while it stretches so high up into the atmosphere that it bursts through into the lower stratosphere.

It is powerful. It packs close to a megaton of energy—about the same as a small thermonuclear bomb. The dog is smart to cower.

It is organized. It is a wet and thunderous thing that sucks warm air from the surface while pumping down cold air from aloft. It feeds on the latent heat in the moisture it takes in. A thunderstorm is like an enormous, natural, self-powered, heavy-duty air conditioner.

Although it originates from random microscopic fluctuations, it works systematically. Specialists say that it is "self-organized." But its system is fashioned out of a strange chaotic fabric, seeming random to human eyes because it is so deeply complicated. Inside the thunderstorm there are wild, turbulent movements of matter in all three states, not to mention the plasma created by its lightning. This tumult is incomprehensibly complex yet structured on every scale, smaller and smaller until one arrives at atomic levels.

What tantalizes scientists about a thunderstorm is this. We have a complete theory for parts of the storm's movements, and the theory captures, in principle, all the detail. Based on experience with other physical theories, such as that for the motions of planets, where computers can be used to forecast specific events decades in advance, we would expect to be able to use the theory for computing too. The catch is that it is too complicated to be computed. Even with a mathematical theory, no supercomputer in the world can capture the imponderable complexity of a thunderstorm, nor is one expected to emerge that can. Furthermore, no experiment can fully test the theory, as important movements and events take place on all scales smaller than any laboratory probes.

This complicated behaviour in a fluid is called **turbulence**. In the past 20 years, publications on topics relating to turbulence-like phenomena have exploded in physics journals. Nonetheless, some very reputable people believe that we will never adequately solve the problem of turbulence. There is no theory that can be mounted onto a computer that will tell us what today's thunderstorm will actually do, even if we may gain some worthy insights with models, which cannot capture that detail completely.

Turbulence is one of the most basic and intractable research problems facing humanity. You can't measure it. You can't compute it. But rain falls because of it. The limits of what we know are closer than many people think. Like Job in the Bible, scientists hear a voice from the whirlwind declaring the mystery of the unknown.

What does the thunderstorm raging overhead have to do with global warming?

At this moment, and at every moment, there are thousands of active thunderstorms in the hot, moist places of the planet. There are tens of millions of them in a year. It should be clear that this great and constant roar of atmospheric air conditioning is an important part of the global energy budget and should figure significantly into any model of the global climate. However, the mighty creature overhead, along with all of its cousins, is too small to show up in even the biggest and grandest global climate models. They are, in the jargon of the field, **sub-grid scale**—computerese for "they fall between the cracks." Not only can we not handle today's thunderstorm, but no such storm ever shows up, even in our very best computer climate models. Thus thunderstorms certainly are not dealt with from first principles in climate models either. However, if they, and other sub-grid scale phenomena, were ignored, climate models would produce absurd results as the collective effects of these phenomena accrue to become global. In climate, sub-grid scale processes are essential, as the grumbler overhead seems to insist on reminding us with his rolling thunder.

This puts climate models into an awkward category in the world of numerical computation. In a normal computer calculation in other fields, one aspires to make the computational grid smaller than all important structures and processes—that is essential to doing a numerical computer calculation from a theory. But this route is barred to climate modellers because important processes take place on all scales. It is the nature of the thing.

People who do serious climate calculations understand this problem and the fundamental scientific dilemma it implies. The only way to produce non-absurd calculations is to make up some ad hoc rules that insert or take away the energy, moisture or momentum as needed to produce sensible behaviour. From the point of view of someone following the basic equations governing the behaviour on the grid, such insertions appear as if conjured out of nothing, while the removals appear as miraculous returns to nothing.

Even so, these made-up rules are not foolishly done. From the collective effects of sub-grid scale phenomena, **parameterizations**—empirical rules that mimic the overall effects of those phenomena fairly closely—are introduced. There is no precise physical theory behind these emperical rules, just some meteorological wisdom and observation. And they are not unique. The use of such parameterizations means the resulting computer calculation procedures are *models* and not computations of basic *theory*. Climate models do not represent a theory for climate. There is no comprehensive scientific theory for climate!

Therefore, forecasting climate change with a model, in lieu of a theory, is a dicey proposition. We don't know if, for example, the thunderstorms will work collectively in a different climate in quite the same way they do now. Parameterizations do not normally conform to the laws of physics, and it is only the laws of physics that are guaranteed not to change with climate. There is no such guarantee for parameterizations.

Engineering models also use parameterizations, but they are tested experimentally for each regime to which the model is applied. No engineer would sign off on a model that was not tested over the conditions it was meant to apply to. But climate forecasting is no engineering problem. We can't test climate models over all of the conditions they are meant to apply to. We have only the current climate on which to test parameterizations. As long as we don't forget this, no great harm is done.

Models remain very useful, but we must not forget that they are only caricatures of nature. Think of the difference between Mickey Mouse and a real mouse. Among other things, Mickey has only four fingers on each hand. The thunderstorm outside is a bit like one of Mickey's missing fingers.

This is not a poor reflection on climate modellers. They are among the very best, working with all their effort, on one of the most difficult scientific problems there is. If the parameterizations steer the rest of the model down the wrong path in the case of a different climate, no one should be too disappointed. No commitments or laws have been broken. That doesn't mean that climate change experiments aren't scientifically useful or interesting. It just means that they are uncertain.

They are uncertain for very fundamental reasons. Their errors will not just be quantitative but qualitative as well. As we will show in Chapter 7, even the notion that carbon dioxide emissions will warm the Earth's surface is up for grabs once the role of parameterizations is understood. Without violating any physical laws, climate models could forecast global cooling! It all depends on how you do the parameterizations.

Knowledgeable modellers are open about this. Among them there is no pretense that matters are otherwise. There is no need to hide these things. We need models, even with their fundamental problems.

Although the idea of global warming due to increasing amounts of carbon dioxide dates back to the 19th century, computer models are now the focus of our thinking about the question. Modern computer models have made people think it might be worthwhile to look for evidence of warming, given that carbon dioxide amounts have been observed to be increasing.

However, somewhere along the line, this straightforward common sense about the uncertainty of models and their relationship to global climate change was pushed aside. It is not exactly clear when the push started, or exactly where, if it started from any particular place, but it is definitely not scientific in nature. It is something foreign that has thrown the international scientific community into chaos and led to an international political storm as governments attempt to reconcile the perceived need for decisive action with an irreducible uncertainty about what we know.

What is being pushed can be best characterized as the **Doctrine of Certainty**. The Doctrine of Certainty (subsequently called the Doctrine) is a collection of now familiar assertions about climate, all of which are to be accepted without question, because, as the Doctrine's supporters say, "The time for questioning is over." The basic not-to-be-questioned assertions of the Doctrine are:

1. The Earth is warming.
2. Warming has already been observed.
3. Humans are causing it.
4. All but a handful of scientists on the fringe believe it.
5. Warming is bad.
6. Action is required immediately.
7. Any action is better than none.
8. Uncertainty only covers the ulterior motives of individuals aiming to stop needed action.
9. Those who defend uncertainty are bad people.

The Doctrine is not true. Each assertion is either manifestly false or the claim to know it is false. We will elaborate below on which is which and why, while clarifying the assertions themselves. However, for now, the question of how such a false doctrine could have come to thoroughly enthrall all parties concerned with this subject—the press, governments, non-governmental organizations, scientific and educational organizations, and corporations—arises.

Certainly no one planned it. There are far too many people and groups with divergent interests. What would be the payoff that these sharply divergent interests could even agree to? Furthermore, the vast majority of those who support the Doctrine within those sectors are undoubtedly well-intentioned individuals with the best interests of our world at heart. Few would enter this field for selfish gain.

The parties work, and each has its own distinct agenda. Within each group there is a range of views on the global warming issue. Shouldn't their efforts therefore lead to a variety of conflicting ideas about warming, instead of a single dominant (false) doctrine? Not necessarily. It is a well-known phenomenon in nonlinear dynamics that when two mechanical clocks, keeping different time, are placed on the same shelf, they can both start to keep the same time. No one adjusts the clocks, they just start to keep the same time due to their interactions through the shelf.

The effect is called *phase locking*, and it means that different systems, having different properties, can start to sing the same tune. All they may need is to be dynamically engaged with one another. This is self-organization in its simplest form. Of course the organized outcome belongs to no one component. If they were human groups, then the outcome need not correspond to the aims of any of the groups. In that sense, the self-organized outcome would be an unintended consequence of their interactions.

This too is a well-known phenomenon in the mathematical theory of games—a field that has recently caught a lot of attention because of the movie *A Beautiful Mind*. A group of individuals share a common goal. Each pursues it with an individual strategy that seems best. Yet, at the end of the game, the group finds itself locked into an outcome that undermines the original aims—an unintended consequence.

The parties concerned with global warming are not independent. They are forced to be fully engaged with each other by larger circumstances. Their relationships are awkward and strained. The press, politicians and scientists, for example, all depend on each other, but regard one another with suspicion. Each group nurtures stereotypes of the others. Everyone knows some of the clichés: Scientists never give straight answers that people can understand or care about, journalists distort and dumb down the truth to sell news to the lowest common denominator, and politicians don't tell the truth.

These clichés suggest a kind of coherence in how the parties interact with each other. The clichés imply strategies: Journalists ignore scientific uncertainty because "scientists never get to the point," or scientists come to think the only way to get the attention of journalists and politicians on their important work is to embellish and exaggerate, because "journalists only want scandal and sensation." While no party is dictating the agenda, the combination of the divergent goals and awkward and untrusting relationships can come together to *self-organize*, just like the thunderstorm outside.

The Doctrine is the product of a sociopolitical thunderstorm. The differences between the parties are the pressure gradients that set up the flow, and the warm, moist air that feeds it is the ambient fear we all can have about an unknown future. The thunderstorm punched through into the sociological stratosphere when the former president of the United States, Bill Clinton, declared, "Global warming is for real."

This is not just a huge fad where scores of people wear strange clothing, dye their hair blue and pierce their bodies until they discover they look silly. The Doctrine has some serious consequences that have already harmed us and will do greater harm than those spinning in the whirlwind seem to realize. When dealing with scientific issues, the reality of nature cannot be set aside on political grounds without paying a price.

On a global scale, not only can inaction lead to negative consequences, but so can inept action. Doctrinaire certainty can lead not only to political confusion, but it can also confuse our thinking and ultimately undermine science—which must strive to be neither doctrinaire nor certain—itself. Make no mistake: On this issue, we need science functioning healthy and free! Not only must we avoid doing harm with our putative remedies, we also need to preserve our clarity of thought and our willingness to learn about the natural world. Much as many would like to believe otherwise, we truly need to learn a great deal more about this world.

This book deals with these issues. It shows how entrenched metaphors, such as the **greenhouse effect**, cloud thinking on the issue of certainty. It questions the construction of the climate change issue in terms of temperature, and it shows the difference between healthy science and the Doctrine in the treatment of climate data. It demonstrates that well-intentioned but misguided action, such as the Kyoto Protocol, can harm many without helping any. And it reminds us of a lesson too often learned the hard way: Good policy on complicated topics is whispered by the still, small voice of reason, sometimes scarcely heard amid the roaring winds of popular opinion.

The thunderstorm has moved on. Apparently it spawned four tornadoes—pretty good for something falling between the cracks. But while the sun has come out, the human storm continues to grow darker and more menacing. And it has already done damage of its own.

2 The Convection of Certainty

In the preface, we promised to show you that most of what you have been hearing about global warming is likely wrong. We shall do that soon enough, but for the moment we must talk about why things have gone wrong. A key theme of this book is the damaging role in science and policymaking of the Doctrine of Certainty: the strange and untenable notion that we are quite certain about things we actually know little about. Before we can move into the specifics of the scientific material, we need to explain how the Doctrine emerged.

This chapter begins by looking at the very human enterprise of science. Like any human undertaking, it must contend with the realities of politics and power, realms that scientists have typically found difficult to navigate. But the story moves beyond scientists, who have largely become bystanders in the global warming story. The real players are identified, and their motions and interactions mapped. We describe how they created and reinforce each other's belief in the Doctrine, and in the process drove out regular science. We call this pattern the **convection** of certainty. We conclude by illustrating the powerful motions of this system with an example straight from the news about changes in Canada's North.

The Scientist's Burden

Civil authorities sometimes rule at odds with the laws of nature. According to legend, King Canute ordered the tide not to come in, but nearly as infamous is the case of the Indiana State Legislature's abortive attempt of 1897 to legislate geometry, which would have made an erroneous value of π state law.

It can be dangerous to be a thinker at odds with official policy about nature. Astronomy was once a particularly perilous subject. Where he was burned to death in 1600 in Rome stands a brooding statue of Giordano Bruno, who had offended official positions by maintaining that stars are suns with planets. Galileo was threatened with instruments of torture and spent his final years under house arrest for his claims. In the Inquisition's judgement against him, he was specifically cited for having held that "the earth moves" and not the sun, "contrary to the Holy Scripture."

But perils to scientific practitioners at the hands of civil authorities are neither limited to astronomy nor to times long past. During the Lysenko episode in the late 1940s, in the Soviet Union, it became dangerous to espouse unofficial views of genetics. The regime preferred to have genetics function according to a politically motivated scheme, rather than according to nature. Some scientists, still not properly remembered, found themselves on the wrong side of this fight. This time they weren't burned; they were shot or died in prison.

Such occurrences are less apparent in recent times, though they certainly still happen. No one is being burned at the stake or shot over unacceptable scientific views these days. The persecutions of scientists take more subtle forms and are masked by contemporary disputes and the political prejudices of our day. They know that while official policy ought to conform to natural law, policymakers would often prefer the reverse. It is all too easy to fall into this trap when intoxicated with a passion to fix the problems of the world. No one is aiming to repeal the law of gravity, at least not at the moment, but scientists know that if there is ever a passion to do so, it would probably be unhealthy to get in the way. Moreover, every scientist has heard stories that make the notion of an attempt to repeal the law of gravity seem not entirely funny.

Thus the paradox: Despite our huge scientific achievements and technological advancements, we live in what the great 20th-century physicist Richard Feynman called an "unscientific age." It is unscientific because there has been a retreat from popular scientific learning for reasons that remain not entirely clear. The mathematician-philosopher Jacob Bronowski explained it as a "loss of nerve."

Every technically trained scientist knows that you don't discuss physics or mathematics in polite company. It's not that people are intellectually incapable of such discussions; it's just considered rude. Of course, some would rightly point out that talking shop on any topic could be considered out of line when not in the shop. Basic knowledge of the natural world is not shop, however, it is what all educated people ought to possess. At least that is what our classical forebears thought when "liberal arts" meant that half of the curriculum should be devoted to science and mathematics.

Today, while far more science and mathematics is known and available for the curriculum, ironically, less and less of it, including some of the greatest human achievements in history, is known. Not only are we not keeping up, we are unlearning what we once knew. Even many educated people are scientifically and mathematically illiterate, because science and mathematics have all but disappeared from the core of a well-rounded education in many places. Many sophisticated and influential people today have a level of scientific and mathematical knowledge that would not compare to that of a monk from the Middle Ages.

Even if we say that science and mathematics are just some kind of specialized "shop" trades not necessary for daily living, what happens when the issue of the day *is* the shop? What if the issue is global climate change—a topic that is unavoidably and irreducibly scientific? Do scientists get to talk shop then?

Alas, no. Instead, the stage is populated by many people whose desire to save the world is deemed an acceptable, even a preferable, substitute for technical understanding. Not only do they feel no obligation to learn what they need to know to discuss the subject properly, it is almost considered bad form to do so. It is a bit like a hockey game where the majority insists on playing without learning to skate or use a hockey stick. Instead of exploring and debating the natural world together, we slip, slide and stumble about, playing politics instead.

This is an old story in the history of science: Being politically active is more comfortable than being scientific for many. But the weather has no political motives. There are no right-wing or left-wing thunderstorms. They do not have the sensibility to understand political intrigues and fears, nor do they care what people predict they will do. They just do what they do. That's all. It is wise to try to relate to them on their own terms and not to deceive yourself into imagining that you can bluster through without doing the necessary homework.

Scientists know that people don't do their homework. And it isn't that everyone should rush out and learn science over everything else, but when science does come to the fore, as it does in everyone's life from time to time, there is no substitute for the hard work it demands.

As often as not, such gaps in knowledge are filled in with fears and insecurities. Without real knowledge, people will act on those fears and make poor choices that leave us worse off. If this is to be avoided, people must be willing to meet scientists at least halfway. Many scientists turn themselves inside out and upside down trying to make their disciplines comprehensible. Their hearts burst with the need to get people to see the incomparable importance of understanding the natural world. They do this knowing full well that it is for naught if people aren't willing to do the necessary intellectual groundwork.

Richard Feynman describes what an encounter with a reporter is often like: "But when it comes to science, for some reason or another, they [reporters] will pat me on the head and explain to dopey me that dopey people aren't going to understand it because he, dope, can't understand it."

Feynman didn't mean that journalists aren't as smart as anyone else, but that they have a different standard for reporting on science than on other subjects. They report on all other varieties of obscure details, but not when the subject is science. Some of them have related their approach to one of the authors, Chris Essex. They called it the "grandmother test." Unless "a grandmother" would understand the story and want to read it, it won't be printed. Chris's reply, that he knows some very bright grandmothers, fell on deaf ears. We have some tolerance for technicalities in business reports, and arcane sports data are always welcome, but ordinary levels of scientific complexity enjoy no such indulgence.

To counter this imbalanced standard for science, the scientist's metaphor machine runs hot on the subject of climate change as scientists try to reach out to close the gap, far past the halfway point. The makeshift metaphors slapped together thus far have come to be prominent substitutes for a proper explanation. The "greenhouse effect" is perhaps the most famous example of what they've come up with in this mug's game. Indeed, it has become so established that many scientists naively imagine that everyone understands that it is an inaccurate metaphor. More will be said about this inconvenient point, and others like it, in the next chapter.

There is a price for these metaphors. Apart from derailing public understanding of the science, simple metaphors get turned back onto the scientists. Why do we need to spend money on complex climate models and arcane theories if the climate problem is so simple? It's hot in a greenhouse, isn't it? Why didn't you scientists get to the point from the beginning? The point of reaching out to explain was to show the public that the scientific thinking matters. What we get instead is the opposite conclusion, because simpler metaphors seem to dispense with all the scientist fuss. This contrary outcome arises when scientists oversimplify things to try to get a serious discussion started—but if they did not do so, there would be no discussion at all. It presents an impossible dilemma for scientists.

The Doctrine of Certainty itself gets turned back on the scientists too. Some scientists would say that we have to express certainty when speaking of climate change so the public will see the importance of climate research, as normal scientific caution doesn't "sell." This results in the argument that if we know enough to be "certain," we don't need any further research and research funding.

It seems like just desserts, but note that no scenario supports basic research on climate change. Without the dire-certainty sales pitch, basic research is treated as unimportant; with it, basic research seems unnecessary. Basic research on climate is more important than ever, but there seems no straightforward way to make that case.

There is still more fallout. What happens if a scientist in another field doesn't recognize these makeshift metaphors for what they are? What if they get serious research results that do not agree with what the metaphor would suggest? What if a metaphor and the Doctrine of Certainty are in agreement, but the research is not? It can lead to a misplaced conviction that the serious research result is wrong.

For individual scientists, the Doctrine of Certainty is ironic. Certainty is anathema to modern science, but its invocation has elevated the science of climate change to global prominence. Many scientists find the Doctrine objectionable on professional grounds, but live with it because it helps to raise an overdue awareness of the importance of understanding climate. Among individual scientists, there is, as always, a wide array of opinions about such matters. There really is no consensus, despite the blunt claims by the Doctrine's supporters.

For the most part, individual scientists are outsiders looking in on their own party. As we will discuss below, **Official Science** holds sway on the subject of global climate change, not individual scientists. Most know that the topic has ceased to be normal science, as anything contrary to the Doctrine has become virtually unpublishable. One distinguished climatologist told Chris that anything contrary to the Doctrine is dismissed as "worse than wrong." You need only look at a recent editorial in *Nature* (discussed later in this chapter) to see what you might be accused of if you were to attempt to publish something out of line.

Scientists are not good politicians. When discussing the case of Galileo, Jacob Bronowski points out that the astronomer thought that the facts would save him from the Inquisition. Scientists continue to this day to have trouble grasping that a correct argument is often not a winning one in politics. It is basic denial. Despite being figuratively burned over and over, they persist in imagining that a tight, logical argument supported by the evidence should win the day. When this tactic fails, as it often does, some befuddled scientists continue in hope, quixotically aiming to set reason against passion the next time.

Many of the rest come to understand—and resent—that they do not do well when there are powerful non-scientific social forces afoot, such as those that have permeated the scientific enterprise concerning global climate change. In such cases, it is best, in their eyes, to stay away from the irrationality of the day by keeping their heads down and their noses clean until the craziness passes. They know what has happened before. It is written in the cowl-shadowed eyes of Bruno's statue in Rome.

The Key Players

In Chapter 1, we likened the current state of play concerning global climate change to the thermal convections that create a thunderstorm. There are five key players awhirl in the global warming debate: public sector politicians, Official Science, environmentalists (private sector politicians), media and industry. This section considers their characteristic motions and maps out how their interactions have combined to produce a situation no one probably intended: an inordinate feeling of certainty over matters that are still unknown. We introduce the players and discuss some of their motivations—what makes them move. Then we speculate how their interactions create a system that is itself in motion and leads to the outcome we call the Doctrine.

Public Sector Politicians

Our government leaders want to make the world a better place through the instruments of government. Politicians and their parties naturally look for good causes that engage people's passions to rally their support. One such cause is environmentalism, with roots in the late 1960s and 70s, which achieved new prominence between 1986 and 1992. Numerous polls taken over the late 1980s showed a sudden leap in concern about environmental quality. Media coverage of environmental matters jumped, North American universities introduced hundreds of new degree programs in "environmental science" in response to student demand, municipalities introduced "blue box" recycling programs, and grocery stores began to carry "green" products like recycled paper and biodegradable cleaners. U.S. President George Bush Sr. declared himself the "environmental president," Margaret Thatcher announced, "I am an environmentalist," prominent U.S. Senator Albert Gore convened special hearings on the threat of climate change, and Canadian Prime Minister Brian Mulroney announced a multi-billion-dollar "Green Plan" affecting all major operations of the government.

It is a paradox that air and water quality was actually improving at this time. Figure 2.1 illustrates this. Between 1970 and the late 1980s,

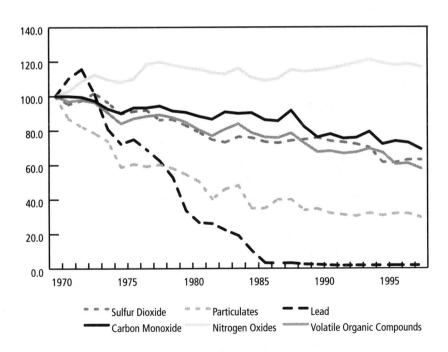

Figure 2.1. Emission indexes for common air contaminants in the U.S., 1970–97. All data from the U.S. Environmental Protection Agency.

economic activity in the U.S., as measured by real gross domestic product, rose 130%. Apart from nitrogen oxides, a smog precursor which rose 20%, then levelled off by the late 1970s, emissions of other major air contaminants *fell* by at least 20%! The picture was similar for many developed countries: Whether looking at air or water quality, most things were in better shape in 1990 than in 1970 in an absolute sense. There was no crisis, according to real, directly measured environmental quality numbers.

What was a world leader to do, faced with a public looking for bold leadership to fix an environmental "crisis," just when the environment appeared to be getting better on its own? Only in politics can this be a dilemma.

It is not clear to what extent world leaders knew that public perceptions were out of line with the facts, but as they say in politics, "perception *is* reality." Everyone believes there is a crisis, therefore good leaders must act, and if action is needed, better make it big. Global warming turned out to be a tempting cause. Anything "global" is inherently more exciting than an ordinary local issue. And "warming" is easy to explain to everyone. There is no arcane scientific mumbo-jumbo to learn or explain—just "hot" or "cold," like the tap in the kitchen. Speeches on the subject can use purple phrases like "international community," "future generations," "bold leadership" and, as always, "new paradigms."

Soon there were the big initiatives leaders seem to love, especially the ones with catchy acronyms. Over the 1990s, we saw the United Nations Intergovernmental Panel on Climate Change (UNIPCC), the United Nations Framework Convention on Climate Change (UNFCCC), the Earth Summit at Rio de Janeiro (UNCED), summits on sustainable development at Toronto and Berlin, six conferences of the parties (COP1–6) to the framework convention, and, ultimately, the Kyoto Protocol on Global Climate Change. Whew! No one could accuse our leaders of inaction.

Dating back to the early 1980s there were still more major committees on the subject of global climate change, each bearing big acronyms. The acronyms are not normally remembered, but they always seem very important before they are forgotten. But while scientists do like their jargon, long acronyms are typically of bureaucratic origin, arising when governments or government organizations create big national and international committees. Their presence is a giveaway that the committee is not of a strictly scientific constitution.

We will frequently have cause to refer to the biggest and most illustrious of these grand scientific committees on climate change: the UNIPCC. We will simply refer to it as the **Big Panel**, or, alternatively, the Panel of the United Nations (PUN). Actually there are three panels, but in this book, when we talk about the Big Panel, we are referring only to Working Group I, which deals with the scientific questions.

In this "unscientific age" such committees can function as a kind of insulator, protecting politicians from the demands of scientific learning on given. Their input is science and the output is executive summaries. There is no need to know the subject or deal with the concerns and doubts of scientists if you have executive summaries.

Official Science and Scientists

Scientists are suspicious of authority in connection with science, and they feel ambivalent about the concept of expertise. On the other hand, the wider society depends on both of these ideas. The result is something we shall call Official Science. Among other places, Official Science lives in the interface between scientists and executive summaries. But we should talk about what Official Science is not before we talk about what it is.

It has nothing to do with the scientific experience and the essential motivations of scientists. The scientific experience always begins and ends as a personal one. The experience is one of patience, puzzles, mysteries and heady excitement as new vistas of understanding open after long personal struggles. Sometimes the insights come at you in a rush, at other times they slip in when your back is turned, leaving the odd sensation that you knew them all along even though you know full well that you didn't. There is absolutely nothing official about this. There is just human ingenuity tested against nature by using hard work.

There is no room in this experience for experts. Experts are middlemen standing between you and the firsthand experience. Firsthand experience is what people care about in science, whether it is through an experiment or an abstract calculation, and this can involve something that no one ever knew, or it can be a case of revisiting what has been done before. If it is your first visit, even though it has been done before, it can be as new and fresh to you as the person who did it the first time. Also, sometimes the person who did it before you got it wrong, or overlooked something, so you may discover something new to everyone after all.

Of course, being realistic, sooner or later some part of what has to be done cannot be firsthand. Secondhand knowledge must be accepted in some part, so scientists must ultimately accept the work of some other scientist as correct. But this is never done at more than a provisional level. Expertise is strictly probationary. Even the most famous and respected scientists can be taken to task and put to the test—if scientific culture is working properly.

Expertise represents something different in the culture of the wider world. There, it means authority in the sense of being authoritarian. Firsthand insight is not so highly valued; instead, authorities define what is true and what is not. If an authority makes a pronouncement, doubting it or suggesting alternatives is not viewed as just seeking the truth, it is taken as a challenge to power. It is seen as arrogant, or provocative, as a test of strength rather than a test of truth. So a political struggle replaces testing an idea. Shrewd politicians know that political power can be enhanced with an air of scientific authority, while scientists know that political power can give science the recognition and resources it needs, not to mention individual gains that may be achieved.

Governments and other such institutions consult and employ people to act as science authorities. The collective voice of these authorities we call here Official Science. These authorities can be government-employed scientists and consultants, scientist-administrators, editors of popular science journals and magazines, and science writers.

Official Science may serve many functions, but it is most important to understand that Official Science is not science. Moreover, those involved with it represent only a minority of people involved with science, and they are not appointed by scientists to speak on their behalf. In fact, as often as not, they are appointed by organizations that have little sympathy for the concerns of scientists.

The lesson of presenting science in a political arena is that one must avoid clashes and pay respect to things that have nothing to do with the science—things that may even be contrary to it. Official Science acts accordingly by trying to strike a mad compromise between the realities of politics and the realities of nature. So while scientists are skeptical of their own work and that of others, Official Science speaks with the simple confidence that good politics requires and journalism demands, but which science abhors.

On January 28, 1986, the U.S. space shuttle *Challenger* lifted off from the Kennedy Space Center and, to the world's horror, exploded in mid-air. The ensuing investigation revealed that, before the disaster, the hands-on engineers responsible for the spacecraft quoted odds of 1 in 100 for failure of the spacecraft, while the administrators (Official Science) instead quoted a figure of 1 in 100,000 for failure. The magnitude of the disparity is as remarkable as its direction is predictable. Official Science often projects confidence that may not be warranted.

Such is the attitude that has infected Official Science on the subject of climate change. While scientists are at odds with each other in many disputes, reflecting wide-ranging opinions, Official Science works to present a united front to its clientele of unschooled politicians and journalists. The "consensus view" on global climate change is unanimous only in the community of Official Science.

The resolve that Official Science has shown on this issue, induced by powerful political forces and the press, has baleful consequences for normal science. A good example of non-governmental Official Science causing damage to normal scientific research can be seen in a recent editorial in the journal *Nature* of July 12, 2001, page 103. A surly denunciation of those who have criticized the Big Panel's latest report accuses them of an "unscrupulous determination to deny the facts" and maligns them as stooges of industry who lack scientific credentials and depend on influence-peddling and media manipulation to make their case:

> Right from the outset, the approach of certain industrial lobby groups in the United States has been to resist, resist and resist again the mounting evidence that the consumption of fossil fuels is producing emissions that change the makeup of the atmosphere and may endanger the future of the planet. The industry groups in question are accustomed to the untrammelled purchase of political power in the United States and have consistently sought to distort the climate change debate for their own purposes.
>
> To this end, they have championed specious scientific findings and worked to establish a bogus scientific debate between

their own "experts"—many of whom are not even atmospheric scientists—and the consensus view of climate researchers. In doing this, they have deliberately set out to take maximum advantage of media gullibility, ensuring that stories on the problem include both "sides" of the debate.

This debate, the *Nature* editorial claims, amounts to a campaign to "confuse and delude the U.S. public on global warming."

It is important to understand that scientists while doubting and debating, also communicate their research findings by publishing in prominent research journals such as *Nature*. Such publications are the lifeblood of modern scientific careers. Despite the peer review system, the editors of such journals have power over what is published. Not only do they have the final say, they can also handpick which authority will give an opinion on a submission to help decide whether it will be published or not.

Imagine now that you are a scientist who has done some work that is not very supportive of global warming. Would you send your manuscript to *Nature* for their decision to publish? Of course not. The editors have stated in black and white what they think of such work—a brazen flouting of the institutional impartiality essential to the health of normal science. If *Nature* was aiming to foster healthy scientific research, rather than show a united front as a part of Official Science, they could not have done a worse thing.

To compound the affair, *Nature* is a leading journal, so other journals and individual scientists, not privy to details on this topic, take a cue from such imbalanced, gratuitous commentary. Before long, someone from the Big Panel will note that there haven't been many papers challenging the Doctrine in a prestigious journal like *Nature* lately, which will then be adduced as further evidence of the high level of scientific certainty on the matter.

Nature is far from the only offender on this issue, and it certainly didn't begin in 2001. People in the field have known for many years that it is an uphill struggle to publish any work seeding doubt about the warming hypothesis, while journals keep on publishing the most incredible things claiming that it is proven.

For example, *Science* has recently been publishing Doctrine-supporting papers that discuss world ocean "warming signals." These are alleged to have a magnitude of—brace yourself—0.1°C over half a century! When was the last time you saw a thermometer, other than a medical one, that could measure more closely than 1°C? What if you had to pick out a 1°C change over areas larger than the major continents, with only a smattering of measuring locations available, where the temperatures everywhere typically vary each day over ten times that amount? Now picture looking for a tenth of that over an interval of time equivalent to that between the birth of your grandparents and you.

In the papers' defence, one can say something about averages over large volumes, but this is a tricky business to do accurately and even more difficult to interpret sensibly. (The problematic physical meaning of an "average temperature change" will be discussed in Chapter 4.) Yet the press pointed to the conclusions of these papers as the "strongest evidence to date" for the Doctrine. There certainly must be a stronger case to make than this.

The Big Panel (PUN) itself represents Official Science. Governments appointed its top officials, such as Robert Watson and Sir John Houghton. Regular scientists do not write the executive summaries that people read. Government-appointed officials, who may not even have significant scientific training, write them. The Big Panel does not call them "executive summaries" but "summaries for policymakers," and they come with their own acronym: SPMs. They may or may not be read by policymakers, but they certainly seem intended for the media. We will call them summaries for policymakers and media, or SPAMs.

Of course scientists do participate in writing actual reports devoted to various aspects of climate and climate change. Sometimes these reports are summaries in their own right; however, these documents share with their much larger, 1,000-page counterparts the sad fact that the executives, or policymakers, as the case may be, don't actually read them.

This is not an insult, although it is lamentable. They are not the sort of documents that a busy, non-scientifically prepared person would be equipped to read. If we lived in a different world, where science was not systematically avoided by so many, this might be otherwise.

For the most part, such reports contain routine meteorological and climatological information. They don't skip straight to the juicy interpretative points the way the SPAMs do.

The big reports are commissioned by the Big Panel at regular intervals and are called "assessment reports." (or the Big Panel ARs). The third of these, appropriately named the "Third Assessment Report," was published in August 2001—seven months after its summary. Curious.

There have been complaints that the SPAMs misrepresent the ARs. In the light of the odds of failure quoted in the Challenger disaster report, this should not come as a surprise. It is precisely in the difference between the SPAMs and the Big Panel's ARs that some of the differences between Official Science and scientists is likely to emerge. Committees can serve to isolate the political world from nasty scientific realities. Legitimate scientists, whether they think that warming is happening or not, are consequently mere spectators in the dynamics of global climate change politics. Official Science has the helm.

The Media

There are no acronyms here. In fact we are not even sure what best to call it. The press? The media? Journalism? News? None of these names seems quite satisfactory, but everyone knows what *it* is. *IT* seems to be about the truth, and indeed one of the most famous newspapers in the late Soviet Union was called *Pravda*, i.e., "truth."

Of course, other newspapers, etc., are more modest, preferring the word "information" over "truth." "Information" seems less ostentatious than "truth," and less likely to cause philosophical disputes. Information is depicted as a kind of substance you can have amounts of. Envision piles of it at a depot, being loaded with shovels, to be carried down an information highway.

Scientists like to think that they are in the truth and information business too, so you would think that there would be a natural rapport between the worlds of journalism and science. However, as we have seen above, this is not the case. There are some notable reasons why. One of the most basic is that, on contentious issues, scientists rarely agree publicly with each other, while journalists rarely disagree.

This is a much bigger difference than it seems. If you have an understanding (stated or not) that you are not going to challenge your professional colleagues in public, it leaves very little to say except what they have already said, or nothing at all. Since the news media exist to say things and do not have the option of not saying things, we end up hearing much the same thing from journalists on any particular topic.

This does not mean that headlines in different newspapers cannot contradict each other. They certainly do. Story coherence on a particular topic takes time to develop, sometimes a few hours, but occasionally days or weeks. And it does not mean that journalists privately agree with each other; they certainly don't. But certain social philosophies can begin to prevail in a profession over time. This is as true for science as for journalism.

But what does not happen for journalists is the sort of followup and criticism of facts that keeps scientists looking over their shoulders when they write. This seems charmingly paradoxical: Scientists seek one truth but often voice many opinions; journalists often speak of many truths while voicing a uniform view.

Once story coherence is established, it sets like concrete. It is composed of a complex network of hackneyed metaphors, hardened clichés and tortured truisms. So, for example, we hear about computer "viruses" rather than programs that people get tricked into running. On climate change, *everything* is "global," even things, like temperature, that are intrinsically local. We hear about "greenhouses," even though greenhouses don't work by the greenhouse effect. We hear about "holes" in the atmosphere. There are no holes.

Just try and correct such things. It is virtually impossible. Write all the letters to the editor you want. Spin out opinion columns all you please. No one is listening. No dialogue will take place. It is like trying to reason with a herd of wildebeest. Understanding doesn't advance as the herd grazes on.

Errors and correction are not a natural part of the journalistic product, and for this reason there is an authoritarian aspect to the news. Correction does take place, but in little sentences as part of retractions, often in reaction to libel complaints. What you will not see is a headline saying that the news item from yesterday was nonsense and here is today's version. Neither will you see a headline pointing out that the headline in the other newspaper is faulty, nor will you see a critique of the other newspaper's editorial decisions.

With politics, the public is, to some extent, skeptical about what the media tells them. On matters of science, however, people do not have the personal knowledge to doubt what they are told. Worse still, science and academia are so cloistered that many scientists from other specialties also lack this knowledge.

This feeds back onto related scientific fields as the media begin to unknowingly set the agenda for science because academics in other specialties attempt to harmonize their research programs with what they perceive is going on through media reports. Most journalists are horrified to think that media stories have taken on such a role. But it has been going on for a long time, and it is a problem that reflects the magnitude and complexity of the modern research environment. However, we can do better if we understand what we do wrong.

Journalism has—for good or ill—a different nature than science. It is driven by very different imperatives. A history professor by the name of C. John Somerville at the University of Florida characterized the most basic imperative in his book *How the News Makes us Dumb*. He observed that news does not happen daily, but daily news must be sold. This commercial pressure gives rise to an addictive stream of trivia called news that bemuses and confuses instead of informing us.

The most effective way to sell daily news is to make it worrisome. Then the public will feel a need to keep posted—by tuning in or buying a newspaper tomorrow. Think, for instance, of how most network newscasts begin. There is dramatic, tension-filled music, a rapid-fire series of clips showing strained faces and burning buildings, and a doom-filled voice announcing the night's top items, always in portentous tones. Why do newscasts begin with tension-building music? Why not, say, a flute arpeggio or some light jazz? You cannot picture it, because these kinds of music put you at ease, but news is intended to make you worried.

The *actual* global warming topic is far too complex for media of this nature to treat properly, but there are aspects of it that fit very well into the commercial imperative of news reportage. If presented in a certain way, it can be used to cultivate alarm about the future. For those who traffic in worry, this makes it a good item to cover once in a while. The treatment of global warming by the media has, from the beginning, been concentrated on coming disasters and gathering doom.

Examples abound. In October 2000, a draft summary of a new Big Panel report was leaked to the British media. The BBC's coverage stressed the worst-case scenarios, took at face value the erroneous suggestion that the projected warming rate had doubled in the past five years, and presented the whole thing on TV with pictures of flooded African villages and inundated Asian coasts. Another example was the "fried egg" issue of *Time* magazine (April 9, 2001), which pictured the earth sizzling in a frying pan and featured in its pages every alarmist claim imaginable. It cannot be supposed that the people who put that issue out actually believed they were communicating science to the general public (after all, what sparked the story was not a new discovery, but a political event: President Bush's decision to reject Kyoto). But it made "great copy," and probably sold well too—though for these authors it was not over easy.

Environmentalists, or Private Sector Politicians
In 1988 the media was hammering the panic button. Attention was focused on the hot, dry summer and wildfires that were burning in the American West that year. Despite observations by many scientists that it was still too soon for warming, given the mainstream climate change scenarios of that time, global warming was repeatedly and willfully offered by the media as the likely explanation. This was concurrent with the United Nation's release of the Brundtland Report on Sustainable Development and the emergence of the "hole" metaphor to describe ozone depletion in the Antarctic. It was a dream come true for environmental activists.

Such activists promote a particular point of view with political methods in the broad sense. While some scientists do associate with them to express their own opinions, environmental activists most definitely do not speak for science. You may have an impression to the contrary because they have been so effective at their particular brand of entrepreneurial politics. They are extra-governmental politicians who form public coalitions in an attempt to sway the media, government politicians, industry and Official Science to a point of view that they strongly believe in. Unfortunately, passionate belief is not a substitute for the truth.

Although this activism has been around for a long time, it dates back to the 1960s in its current form. The events of 1988 were a tonic for the movement: Environmental lobby groups got thousands of new members and instant, newfound respectability. They broke through a kind of social barrier from the periphery, where they had been lurking with other lobby groups, into a special sort of mainstream. No longer were they dismissed as ponytailed eco-nuts; they were invited to address governments and sit as respected members on very serious UN committees, including the Big Panel.

You knew they had arrived within the same hallowed bureaucratic place as Official Science when they stopped being referred to as activists and lobbyists and were elevated to their very own special acronym: NGOs (non-governmental organizations). We will call them ELGs (environmental lobby groups).

This bonanza did not outlast the 1988 heat for long. The press and public and the weather can be fickle. Cool weather and the recession in 1991–92 hit them hard, melting away much of their support as interest turned to unemployment rather than pollution. Memberships and revenue fell. People still cared about the environment, but it wasn't hot any longer and the press had extracted their last drop of excitement from the story.

But looking for an issue to keep their profile and fundraising on track, many ELGs focused their efforts on global warming. ELGs survive by public fundraising, and global warming is a very effective topic to promote: It is big, baffling and ominous. It is so complex you can say just about anything and rarely face a technical challenge that will count publicly. People have the impression that the scary stories have widespread scientific support, even if they aren't sure what the nature of that support is or what the issues actually are.

Furthermore, political buttons can be pushed to distract attention from the scientific issues. There is, for instance, just enough of a connection to industry to insinuate plots by big multinationals and make them stick for a ready constituency. Global warming is tailor-made for the ELGs, and they have made the most of it.

Industry
On one level, industry is simple to understand. Firms exist to make money, and that sums up their motives. But different types of firms make money in different ways. The coal and oil industries make money by producing fossil fuels and selling them. Since these fuels produce CO_2 emissions, the coal and oil companies have a lot to lose if governments enact policies to fight global warming by limiting coal and oil consumption. You might think, then, that these firms would be out fighting against environmental groups and trying to muffle debate on climate change. Surprisingly, they are not. The majority of fossil fuel firms are either silent about climate change or have been subsidizing environmentalist organizations that fight *against* global warming. They have even given money to groups who warn against a big business conspiracy! To understand this

seemingly paradoxical behaviour, we need to first describe the motions of the main players in their recent interactions over climate change.

The Motions of the System

The convection of certainty is driven by the interaction between two groups: Official Science and governments. The energy of the system is their stated commitment to the Doctrine, or the level of certainty they claim to have concerning human culpability for global warming. The system begins in the lower left corner of Figure 2.2, at the sun symbol. This was the situation up to the mid-1980s, when politicians and Official Science alike claimed little or no certainty on global warming. Once the system is set in motion, the arrows show the direction of movement. When politicians claim little certainty, Official Science has only a small tendency to increase its level of certainty (arrow A). But when politicians declare a high degree of certainty, Official Science responds by rapidly increasing its certainty level (real or apparent) as well (arrow B). When Official Science professes low certainty, there is little pressure for

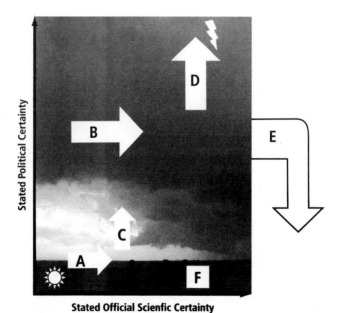

Figure 2.2. The convection of certainty between politicians and Official Science.

politicians to raise theirs (arrow C). But when Official Science professes "99%" certainty, the pressure is enormous for politicians to likewise raise their stated level of certainty (arrow D). The system punches through into the storm level at the top right, at which point the sincere convictions of all participants maintain the high level of certainty we now see proclaimed by politicians and Official Science alike.

Horizontal Movements

The effective role of the media is to amplify the horizontal arrows, A and B, by overstating the degree of scientific consensus and making dissent more difficult. They sample the information given out by scientists, pick out the scary stories and make them sound more worrisome. In the process, they create an impression of uniformity in scientific opinion.

There were hundreds of articles of professional, peer-reviewed research on global warming published in August 2000, but it was all ignored. Two unrefereed pieces of "science" made the news, though, and not coincidentally frightened the public.

One story concerned an oceanographer, a paleontologist and a group of tourists aboard the Russian icebreaker *Yamal* who saw open water at the North Pole. After some casual remarks to a reporter, it was hailed in the press as "proof" of a global warming disaster. The *New York Times* and many others claimed that the North Pole had melted *for the first time in 50 million years*. Shortly thereafter some clarifications emerged: It turns out that open leads of water are quite common in high summer up in the Arctic, where average temperatures reach about 0°C and the thin ice often breaks up. There is nothing unusual about open water in and around the pole in August. If the media had troubled, for instance, to talk to companies that organize Arctic tours (e.g., www.libfind.unl.edu/alumni/tourin/ north_pole.htm), they would have learned that vessel captains always follow open water channels to the pole as far as possible. This particular Web site even promises travellers that, once there, "If you are so inclined, take the 'polar plunge,' a quick dip in the Arctic Ocean." At least tour companies know what the high Arctic is like in summer.

The only meltdown was the credibility of the journalists, but was there any medium by which their claims could be criticized? We regret to say that what you are reading is it. There will be no shiny, equally high-profile national magazines pointing out how foolish they were and how people were misled and upset for no good reason. Worry is always pumped up, but no comparable forum pulls it back down when it is unwarranted.

A week later, the WWF—the World Wildlife Fund—released a report in Canada, warning of massive destruction of ecosystems and habitat over the coming decades as a result of global warming. It warned that 46% of Canada's existing habitat would be destroyed by the effects of rising carbon dioxide levels, turning it into "wastelands devoid of life," with devastating consequences for animal and plant life everywhere. But other places would fare even worse than Canada: Iceland, for instance, would lose 81% of its "life-sustaining habitats." The report received front page coverage in *The Globe and Mail* on August 31, 2000.

This entire story was traced to a computer modell used by a University of Toronto ecologist. One might ask some tiresome questions. How were climate changes and ecosystem changes modelled in it? What rules were fed into the computer? How well did these rules represent things that are known to be true? If the model was set up for the conditions of 1960, how well would it predict the world today? These and others are routine questions to ask of any modelling experiment. It is easy to construct computer models to bring on an apocalypse in 10,000 ways. Kids do it all the time on computer video games. The trick is to come up with a model that tells you something that is true—which is why tiresome questions are necessary with models.

If we lived in a scientific society, these would be the questions the reporter would begin with, and the readers would demand they be asked. Certainly the WWF wouldn't be able to issue reports without going into such details. Certainly one computer modelling foray into this field could not be regarded as the number one national news of the day, nor would it be the stuff on which political positions and expensive policy decisions rest.

Unfortunately it seems to be exactly the material from which editors want to make headlines. Clearly the *selection* of what is newsworthy is key to the process. One of the most important pieces of climate research published in 2001 was an article in the *Bulletin of the American Meteorological Society* of March 2001 by Richard Lindzen and his coauthors. The American Meteorological Society considered it so important they posted a news release about it on their Web site for over a month. In the study, a team of scientists from MIT and NASA discovered that when the surface of the Pacific Ocean warms slightly, changes in cloud formation allow atmospheric infrared radiation to escape easily to space, offsetting the warming effect of "greenhouse" gases. The major media, who were fixated that month on President Bush's decision to reject the Kyoto Protocol, did not pick up this item.

The connection between the two stories would seem obvious: New results suggest the science of global warming is uncertain, therefore we should hold off on expensive policy initiatives. But if you look at the media as purveyors of *worry*, their reaction makes more sense. We were told that scientists are *certain* about the global warming crisis, but an intransigent president is rejecting the proposed solution, preferring to gamble with the earth's future. The "fearmonger" model for media behaviour makes sense of myriad episodes a "journalism" model cannot explain.

Another selection strategy of the media is the lack of attention paid to scientists who dispute the Big Panel position. Sometimes they are even the subject of aggressive skepticism. Many scientists have taken steps to put their opposition to the Doctrine into the public record. The Leipzig Declaration of 1997 reads, in part:

> We believe that the dire predictions of a future warming have not been validated by the historic climate record, which appears to be dominated by natural fluctuations, showing both warming and cooling. These predictions are based on nothing more than theoretical models and cannot be relied on to construct far-reaching policies. As the debate unfolds, it has become increasingly clear that—contrary to the conventional wisdom—there does not exist today a general scientific consensus about the importance of greenhouse warming from rising levels of carbon dioxide.

This statement was signed by more than 100 atmospheric scientists and climatologists from around the world, including, among others, the editor of *Climate Research*, the editor of *Atmospheric Research*, a member of the Nobel Prize selection committee in physics, the chair of the (U.S.) National Research Council Carbon Dioxide Assessment Committee, a former president of the U.S. National Academy of Sciences and a former director of the U.S. Weather Satellite Service. (The full list of signatories can be found at www.sepp.org/LDsigs.html.) Anyone reading the list of signatories would realize that dissent from the Big Panel's conclusions is not limited to a few marginal skeptics, as has been charged.

Equally striking is the Petition Project (www.oism.org/pproject/), based on a critique of global warming research originally distributed over the Internet and subsequently published in *Climate Research*. The paper states:

> While CO_2 has increased substantially, the large temperature increase predicted by the IPCC [PUN] models has not occurred.... Not only has the global warming hypothesis failed the experimental test; it is theoretically flawed as well. It can reasonably be argued that cooling from negative physical and biological feedbacks to [greenhouse gases] will nullify the initial temperature increase.... Human use of coal, oil and natural gas has not measurably warmed the atmosphere, and the extrapolation of current trends shows that it will not significantly do so in the foreseeable future. It does, however, release CO_2, which accelerates the growth rates of plants and permits plants to grow in drier regions. Animal life, which depends on plants, also flourishes.... We are living in an increasingly lush environment of plants and animals as a result of the CO_2 increase.

The accompanying petition states, in part, "There is no convincing scientific evidence that human release of carbon dioxide, methane, or other greenhouse gases is causing or will, in the foreseeable future, cause catastrophic heating of the Earth's atmosphere and disruption of the Earth's climate."

To date, there are over 19,000 professional signatories from the U.S., including 2,660 physicists, geophysicists, climatologists, meteorologists, oceanographers and environmental scientists, and 5,017 scientists from other disciplines. However, the unproven claim that most scientists agree with the few dozen people who write the Big Panel reports continues to be routinely put forward by the media. This, in turn, gets picked up and repeated by the Big Panel itself.

A story in the *Manchester Guardian* of April 6, 2001, quoted PUN co-chair Robert Watson confidently downplaying the possibility of uncertainty on the part of scientists: "[Watson] dismissed any suggestion that there was a significant split among scientists over whether climate change is occurring, and whether humans are causing it. He said: 'It's not even 80-20 or 90-10. I personally believe it's something like 98-2 or 99-1.'" Likewise, at a news conference a few months earlier, the other Big Panel co-chair, Sir John Houghton, said "I think there are very few scientists who'd disagree with the IPCC [PUN]. And most of those who do disagree have not published much." [1]

These slights against the credentials of those who dissent creates peer pressure for those who disagree with the Big Panel to keep quiet. In causing the dissenters to censor themselves, it exacerbates a false appearance of unanimity among regular and Official Science. Sometimes the attacks on dissenters by politicians and Official Science have been unusually forceful. During the Clinton-Gore administration, during an interview on Washington's WAMU-FM (July 21, 1997), Interior Secretary Bruce Babbitt was asked why some reputable scientists did not support the administration's view of global warming. His answer was:

> [It's] an unhappy fact that the oil companies and the coal companies in the United States have joined in a conspiracy to hire pseudo-scientists to deny the facts . . . the energy companies need to be called to account because what they are doing is un-American in the most basic sense. They are compromising our future by misrepresenting the facts by suborning scientists onto their payrolls and attempting to mislead the American people.

The force in an arrow like B in Figure 2.2 can approach hurricane strength. It is driven by a high level of political certainty. Regular scientists notice these things, and quietly back off from the controversy through self-censorship. Bruno's fate haunts us still.

Vertical Movements

Environmental lobby groups contribute to the storm by amplifying the vertical arrows, C and D. When official scientific certainty was low, there was little incentive for politicians to raise their stated level of certainty. Nevertheless some did so to bolster their credentials with the growing environmental movement, most notably then U.S. Senator Al Gore. He convened hearings in 1988 to investigate global warming, and it was at one of these sessions that NASA scientist James Hansen declared that he was "99%" certain the Earth had warmed and that greenhouse warming was occurring. Things began to snowball quickly after that. MIT meteorologist Richard Lindzen described in a Cato Institute essay some of what followed:

> As most scientists concerned with climate, I was eager to stay out of what seemed like a public circus. But in the summer of 1988 Lester Lave, a professor of economics at Carnegie Mellon University, wrote to me about being dismissed from a Senate hearing for suggesting that the issue of global warming was scientifically controversial. I assured him that the issue was not only controversial but also unlikely. In the winter of 1989 Reginald Newell, a professor of meteorology at the Massachusetts Institute of Technology, lost National Science Foundation funding for data analyses that were failing to show net warming over the past century. Reviewers suggested that his results were dangerous to humanity. In the spring of 1989 I was an invited participant at a global warming symposium at Tufts University. I was the only scientist among a panel of environmentalists. There were strident calls for immediate action and ample expressions of impatience with science. Claudine Schneider, then a congressman from Rhode Island, acknowledged that "scientists may disagree, but we can hear Mother Earth, and she is crying." It seemed clear to me that a very dangerous situation was arising, and the danger was not of "global warming" itself.

The push for decisive political action—even amidst very low scientific certainty—came not only from ELGs but also from Official Science in the Big Panel. In the early days of the panel (1988 to 1991), Official Science argued that the politicians pushing the issue needed a neat, clear conclusion to motivate policy. But the regular scientists insisted that simple conclusions were not available. The co-chair of the Big Panel, Sir John Houghton, explains how this dilemma got resolved and the political result:

During the preparation of the reports, a considerable part of the debate amongst the scientists has centred on just how much can be said about the likely climate change next century. Particularly to begin with, some felt that the uncertainties were such that scientists should refrain from making any estimates or predictions for the future. However, it soon became clear that the responsibility of scientists to convey the best possible information could not be discharged without making estimates of the most likely magnitude of the change next century coupled with clear statements of our assumptions and the level of uncertainty in the estimates. Weather forecasters have a similar, although much more short-term responsibility. Even though they may feel uncertain about tomorrow's weather, they cannot refuse to make a forecast.... It has often been commented that without the clear message which came from the world's scientists, orchestrated by the IPCC [PUN], the world's leaders would never have agreed to sign the [Rio] Climate Convention.[2]

This astounding statement by Sir John is the very anatomy of Official Science at work between scientists and politicians. It gathers in normal science in all its tumultuous reality: open debate, dissension and a refusal to make definitive claims where none are to be had. Then it trots off to Capitol Hill or Number 10 Downing Street with a serene and smiling certainty. Debate and dissent are extruded into a "clear message which came from the world's scientists, orchestrated by" the Big Panel. The reason for this orchestration is clearly expressed: Without it, "the world's leaders would never have agreed to sign" a treaty. If things were as they should be, leaders would want a treaty because they observe that scientists have lined up in unison. What happens instead is that Official Science "orchestrates" a consensus because leaders want to make a treaty.

How was this orchestration achieved? The only way we know to accomplish this with a bunch of independent-minded scientists is to arrange things so that those who play along with the tune are most likely to make it into the orchestra. And what do climate forecasts decades into the future have to do with weather forecasting anyway? Weather forecasters indeed do not decline to make daily forecasts. But they do decline to make multi-decade or century forecasts as a basis for major policy decisions.

This twisted logic can only make sense in the light of the discomfort society has with science and scientific ways. Official Science exists for science as a tonic against the media and politicians, but it can also send a message that undermines science itself. Does anyone think the media and politicians don't see this for what it is? Isn't it clear that they are only going along with it because it suits their own agendas? And when it stops suiting them, what will Official Science have gained in exchange for giving up scientific integrity?

ELGs have forced the level of declared political certainty upwards by supplying selective information to the public. Global warming is always discussed by such groups using Doctrine language: We are certain; only a handful of unimportant skeptics doubt the science; action is needed now. Greenpeace says: "The overwhelming conclusion is that there will be significant negative impacts on many natural and human systems even if the tuning and exact character of these impacts cannot be predicted with certainty" (www.greenpeace.org/~climate/, March 2000). The Sierra Club warns: "Increased flooding, storms and agricultural losses could devastate our economy. Plants and animals that cannot adapt to new conditions will become extinct. The world's leading scientists project that during our children's lifetimes, global warming will raise the average temperature of the planet by 2.7 to 11 degrees Fahrenheit." (www.sierraclub.org/globalwarming/dangerousexperiment/, July 2001).

Prominent Canadian environmental activist David Suzuki writes frequently about global warming. On his Web site, he states:

> While warmer temperatures may sound appealing to some
> Canadians, climate change carries a potent threat—extremes of
> hot and cold weather, floods, drought, and destructive storms.
> Climate change endangers human health, and the health of our
> ecosystems. It jeopardizes agriculture, forestry, supplies of fresh
> water and the survival of some animal species.
> (www.davidsuzuki.org, September 2000).

The accompanying document doesn't supply any sources for these claims, but that doesn't matter because this is what "scientists" are saying.

In February 2001, Dr. Suzuki was interviewed in the University of Guelph student newspaper (*The Ontarion*) on the problem of global warming. Among other things, he said:

> It's not environmentalists that are driving this [concern over climate change], it's scientists, the leading scientists of the world are telling us, we are in global warming, we have got to take drastic action and there are solutions available.

Thus are ELGish spells cast on politicians. It is often surprising to those not familiar with the workings of government how much policymakers turn to activists for advice. This is certainly true in the area of environmental policy, and the Big Panel is no exception. Indeed, one of the lead authors of the scientific report (Working Group I) was Michael Oppenheimer, chief scientist of the activist group Environmental Defense (formerly the Environmental Defense Fund, or EDF). It is noteworthy that the organization for which he is chief scientific spokesperson has been committed to the global warming Doctrine for over a decade—before the research even began. In its annual report in 1991, back when even the Big Panel said there was little agreement on the subject, the EDF said: "While it is still uncertain how quickly these 'greenhouse' gases will affect future climate, there is a scientific consensus that global warming will result. Wildlife, agriculture and human societies could face new hardships as climate zones shift and sea levels rise."[3]

The best illustration of the role of ELGs in the storm is their reaction to Bush's decision to reject the Kyoto Protocol. There are many good reasons to reject this agreement: As we explain later, even those who truly believe in global warming are hard pressed to mount a rational argument on its behalf. But Bush's move prompted such an outpouring of rage that other politicians were quick to distance themselves, even those who privately agreed with the decision. And the U.S. administration went to some pains to claim that it was still certain global warming is a problem—they only had some procedural objections to Kyoto. The high degree of official scientific certainty makes it easy for activists to expand arrow D, creating powerful pressure for politicians to declare their certainty on global warming.

The rising column of political certainty in turn feeds the horizontal thrust of Official Science certainty through the mechanism of research funding. Government agencies whose job is to validate the certainty pronouncements of cabinet ministers will find subtle ways to direct the research agenda and pre-filter the results. For example, the U.S. government[4] recently announced a US$300,000 project to produce "Location-Specific Assessments of the Impacts of Climate Change and Variability on Aquatic Ecosystems and Water Quality." These will be tricky to produce, since there is no factual basis on which to do the studies. There are no reliable regional climate change simulations, and no modellers in the world claim to be able to produce accurate long-range climate change forecasts on a local scale with their models. But none of this matters, because the desired conclusions of the studies are clearly spelled out in the announcement, which explains the guidelines reviewers will use in deciding whether to fund an application. Applicants must begin by picking a threatening aspect of climate, then:

> Reviewers will consider the significance of the threat posed by climate change and variability, on (a) water quality, and/or (b) aquatic ecosystems and the importance of the threatened resource. (Applicants should specifically identify the threat, and explain the importance of the resource.) Specifically, applicants should discuss:
>
> (a) Water Quality: assess how climate change and variability might affect the ability of the entity (e.g., state, tribe, municipality, county, or territory) to meet water quality standards for drinking water, surface water, and/or groundwater.
> (b) Aquatic Ecosystems: assess how climate change and variability might affect aquatic ecosystems. The proposal should identify aspects of aquatic ecosystems that are potentially vulnerable to climate change and variability and are of local concern. The concern could arise from regulatory responsibilities (e.g., responsibilities of a state, tribal or other local government to protect aquatic species, habitats, and organisms) or from other stakeholder interests (e.g., economic, recreational, spiritual, or cultural).

Presumably, those who think it possible climate change will be *beneficial* to a local community will not bother applying. And it is a safe bet that the studies resulting from this competition will "discover," among other things, that *climate change may threaten the ability of local governments to meet water quality standards for drinking water, surface water, and/or groundwater* and will find *local aquatic ecosystems may be vulnerable to climate change and variability.* We can picture the headlines already.

And when these studies appear, the press will tell the public to worry about them, Official Science will quote them to illustrate the certainty of the problem and the need for more research, politicians will meet to "do something" about them, and the Doctrine will be reinforced yet again.

This is a positive feedback loop driving up certainty. And is not by any means the only one.

The above-mentioned program is expensive enough—$300,000 is nothing to sneeze at. But compared to other government initiatives it is chump change. Canada devotes $50 million per year to a special research program called the Climate Change Action Fund (CCAF). This invites projects under several headings, including Science, Impacts and Adaptation, Public Education and Outreach, and others. If you are interested in applying, the Web site is http://climatechange.gc.ca/english/actions/action_fund/index.shtml. But be forewarned: Under no circumstances will projects that challenge the Doctrine be funded. The research they have in mind "is intended to support early actions to reduce greenhouse gas emissions" and to increase understanding of the benefits of the Kyoto Protocol. As for basic scientific work, under the heading Science, Impacts and Adaptation, they say: "The CCAF will continue to improve our climate change knowledge, giving us the tools to provide better scientific advice on greenhouse gas emissions reductions." That the gases are harmful and need to be reduced is beyond discussion. The only purpose of science now is to advise on how to do it.

As for Public Education and Outreach, project proposals are welcome that serve "to increase Canadians' understanding of climate change and encourage them to reduce their greenhouse gas emissions." With the intended outcome so firmly fixed ahead of time, why bother with the research, other than to feed back onto itself?

Incidentally, the protest is occasionally made that when industry is involved in funding climate research the results are automatically tainted.

People making this argument should attend to the fact that government-funded research has become the real problem in climate change science.

The motions outlined in Figure 2.2 have lifted considerable political and scientific mass up to the top right-hand corner, where storm activity forms. Arrow E represents the precipitation that has resulted. This is regular scientists dropping out of the whole debate. Despite what the Big Panel and Official Science claim, the majority of regular scientists in the world want nothing to do with this debate anymore. They are keeping their heads down and dropping out. They hear comments like Secretary Babbitt's and read editorials like the one in *Nature*, and they recognize that they are better off not getting involved.

Finally, what *about* industry? Some firms have gotten involved by pressing for restrictions on carbon dioxide emissions. For instance, when the Australian Parliament held an inquiry into the Kyoto Protocol, the Orbital Engine Corporation of Balcatta, Western Australia, made a submission stating they support early ratification of the agreement in its current form. This is not because they have an opinion one way or the other about the radiative effects of CO_2 on the general circulation of the atmosphere, but because they make direct fuel injection motors that operate at high efficiency, and they figure that legislation requiring emission reductions will increase demand for their products. We will discuss in a later chapter where the fallacy lies in counting increased demand for the ingenious products made by the Orbital Engine Corp. as a social benefit of climate change policy. For now, we need only point out that very few firms are in the position of seeing a benefit from a policy like Kyoto.

Most of industry is in the position denoted by the rectangle F. If the storm really breaks, in the form of a ratified and implemented Kyoto Protocol, they will be in the line of fire. Because of that, many people automatically assume industry folks should be ignored. The *Nature* editorial above is a case in point. And any researchers who have been funded in the past by the energy industry are immediately suspected of producing tainted research. Industry has been neutralized by this intense suspicion. If they "support" the Doctrine, well and good. If they oppose it, that just shows they can't be trusted.

Some, however, cannot avoid being drawn into the debate. Most interest has been focused on the big oil and gas firms like Shell, Suncor and British Petroleum. You might think they are fighting hammer and tongs against the global warming scare. Surprisingly, they are not. Instead, they are subsidizing environmentalists who promote the Kyoto Protocol. BP, Shell and Suncor, for instance, donate money to Environmental Defense to support a program called Partnership for Climate Action (see *Chemical and Engineering News*, October 23, 2000.) Their Canadian operations have funded work by global warming ELG the Pembina Institute (see www.climatechangesolutions.com). And the pro-Kyoto lobbyists at the Pew Center in the U.S. boast an impressive list of corporate sponsors from the oil and gas sector, including BP, Shell and (until recently) Enron.

What are these firms doing? Consider the behaviour of stores and restaurants in neighbourhoods dominated by protection rackets. They pay money to a bunch of gangsters in exchange for "security," i.e., so the same bunch won't come back to burn down the store and beat up the owner. And when police try to investigate, the store owners rarely finger the extortionists: They may even praise their good character and deny that they have a problem with them (remember the old Monty Python skit about Doug and Dinsdale Piranha?) They are too scared to co-operate with the police.

Energy companies pay money to moderate environmental groups to purchase goodwill from the larger green movement, because they are scared of vandalism and boycotts. They post earnest messages on their Web sites about how committed they are to finding solutions to global warming and they run ad campaigns showing how groovy and world-conscious their scientists are. As a result, the moderate ELGs publicly congratulate these "progressive" firms and their leaders for their commendable stance. This is generally enough to prevent the company from being targeted for direct action by the more radical wing of the environmental movement.

One firm that has conspicuously decided to not pay protection money is Exxon. The president of its Imperial Oil division, Robert Peterson, made a speech in April 2001 reiterating the company's rejection of the Kyoto Protocol and referred approvingly to an article in the company's newsletter expressing skepticism about global warming science and the claim of scientific consensus. The result was predictable. In May 2001, a network of activists (www.pressurepoint.org) supported by celebrities Bianca Jagger and Annie Lennox held a press conference in London announcing the start of a worldwide boycott campaign against Exxon.

They are now backed by Greenpeace, Friends of the Earth and even some members of the European Parliament.

To summarize, Figure 2.2 maps the convection of certainty between Official Science and politicians. As the stated level of certainty in one group grows, the pressure builds on the others to "be certain" too. The flow is amplified by the media and environmental activists. It leads inexorably upwards to the top right corner, where politicians are certain and Official Science is certain, and no one dare say otherwise.

From Polar Bears to Kyoto: A Microclimate Case Study

We illustrate the power of this system with a case study from Canada's North. The background to this story is summed up in Figure 2.3, which shows average Arctic air temperature anomalies from 1970 to the present. "Anomalies" are statistician-speak for differences from an **average**, in this case the local average temperature between 1961 and 1990. Climate models predict "greenhouse" warming in the Arctic will be stronger than normal, and temperatures for the past two decades do seem to be

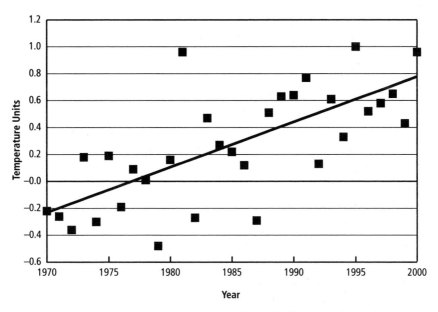

Figure 2.3. Average air temperature anomalies (differences from the mean) in the Arctic, as computed by the Hadley Centre in the UK, 1970 to 2000.

trending upward. The Big Panel frequently refers to the Arctic as a region where the signs of global warming are already evident. Figure 2.3 and the warnings from Official Science have planted the seed of a belief in unusual Arctic warming.

With that in mind, the Canadian government funded the International Institute of Sustainable Development to send researchers up to Sachs Harbour on Banks Island, in the western Canadian Arctic, in 2000 to interview Inuit elders about their observations of climate change. The elders reported that because the ice breaks up more quickly in the spring, they cannot hunt for seal on the ice floes as easily as they used to. Neither can the polar bears, who are going hungry as a result. Other changes are occurring that are interpreted as unprecedented local climate warming. The conclusion? After thousands of years of survival, the traditional Inuit lifestyle is now threatened. All this was attributed to global warming in general, and fossil fuel use in particular—a very long chain of loose reasoning indeed.

It made for several worrisome news items on the CBC in November 2000, and a video was made, which is available from the ELG, EcoAction (www.ecoaction.ca), run by the Pembina Institute mentioned above as a sort of sub-ELG. In November 2000, the makers of the video took it to show to a Big Panel conference convened in Ottawa to discuss how to reduce carbon dioxide emissions. The theory, apparently, is that a non-random sample of Inuit recollections about recent springtime temperatures around Sachs Harbour reveals long-term changes in the world's climate system that are unnatural and deleterious, and that reducing industrial CO_2 emissions would "help."

The next dispatch from the north was on December 12, 2000, when the CBC carried a worrisome story that polar bears around Churchill, Manitoba, are threatened by global warming. Apparently the ice is disappearing in Hudson Bay. Shortly after that, in March 2001, some federally funded scientists went on a field trip to look for more evidence of global warming around Churchill. The media again appeared on the scene. In a CBC broadcast on March 15, the director of the project warned: "We're starting to warm up very, very fast...winters will cease to exist." He also warned that Hudson Bay will be ice-free within 50 years.

The concept was now in full motion: Sea ice is melting, polar bears are starving and the traditional lifestyle of aboriginal peoples is threatened. It is a bit of a leap to conclude that this is the result of fossil fuel use as opposed to, say, the normal cycles of warming and cooling (called the Arctic Oscillation) that have been in a warming phase since the late 1970s, not to mention the usual predator-prey cycles that affect all wild species.

But the interactions among the politicians (who funded the research, told the scientists what to look for and gave the makers of the video an audience at an international meeting), Official Science (who gave the media worrisome stories about polar bears around Churchill and thin ice around Sachs Harbour) and the media (who conveyed all this to the public, heightening public alarm and putting pressure on politicians to "do" something) forms an ascending spiral that reaches toward the stratosphere of certainty.

Around this time, in March of 2001, the University of Western Ontario invited a distinguished U.S. scientist, Professor S. Fred Singer, to deliver the 2001 Nerenberg Lectures on science to a broad audience. Professor Singer is an atmospheric physicist emeritus at the University of Virginia, former director of the U.S. Weather Satellite Service and president of the Science and Environmental Policy Project. He offered to do a talk on his favourite subject, global warming.

It was to be fairly simple. There are four main ways scientists measure temperatures changes: surface thermometers, satellites, balloon-borne instruments and temperature proxies such as tree-ring widths. Only the averaged surface thermometers show a net increasing trend since 1940. The other three do not. Normal science would make this disagreement the centre of discussion and research. He advised that until such fundamental questions are resolved, policy on global warming should be confined to low-cost domestic options, rather than very costly international treaties like the Kyoto Protocol.

Professor Singer has presented this talk at many universities and research institutes in the U.S. and Europe. We thought, naturally, that Canada's Environment Minister, David Anderson, would be interested in hearing it. After all, it is his responsibility to decide what Canada should do about global warming, and it seems like a good idea to decide if there really is a global warming problem. So we sent him an invitation and a description of what the lecture would be about.

Mr. Anderson did not come, but he did reply to our invitation some time later. His letter, reproduced in Figure 2.4, reveals the altitude of the

Minister of the Environment   Ministre de l'Environnement

Ottawa, Canada K1A 0H3

APR 2 3 2001

Dr. Ross McKitrick
Assistant Professor
University of Guelph
College of Social and Applied Human Sciences
Department of Economics
Guelph ON N1G 2W1

Dear Dr. McKitrick:

Thank you for your letter of January 26, inviting me to attend the 2001 Nerenberg Lecture at the University of Guelph. I apologize for the delay in responding.

I am fully aware of Professor Singer's skepticism about the science of climate change. There are uncertainties in climate science – as, indeed, there are in all areas of science. However, the Government of Canada is extremely concerned about climate change and that there is more than enough evidence to warrant aggressive action to deal with the problem. Our North is already experiencing severe impacts. Sea ice is melting, polar bears are starving and the traditional lifestyle of Aboriginal peoples is threatened. The recent report of the Intergovernmental Panel on Climate Change stated that there is conclusive proof that the climate has changed and that this is the result of human activities.

Unless we take action to reduce emissions, the impacts will only worsen. I am confident that progress will be made, one way or another, towards the global solution that is essential to deal with the climate change challenge. I am committed to ensuring that Canada does its part.

I appreciate your invitation to attend the Nerenberg Lecture and receive a briefing from Professor Singer, Professor Essex, and yourself. I regret, however, that my schedule did not permit me to take advantage of your kind offer.

Yours sincerely,

David Anderson, P.C., M.P.

Canadä

Figure 2.4. Letter to the authors from David Anderson, federal Environment Minister.

Canadian government's sense of certainty: We have "more than enough evidence to warrant aggressive action," and those who profess uncertainty are just being "skeptical."

Thus far, we have only shown that the minister treats the CBC news as a scientific briefing, not that there is anything wrong with believing the polar bears near Churchill are starving or the Inuit lifestyle is being lost. So let's turn to the question that would, to a regular scientist, be an obvious starting point: Is it getting unusually warm in the Arctic in general or around Churchill, Manitoba?

Because of the simplistic way the issue is framed, the data are typically discussed as simple averages over large regions. Over the next two chapters, we will show you why this is scientifically wrong, but for the moment we will engage the debate on that level. Over the past 20 years, temperatures have indeed gone up around Banks Island, where Sachs Harbour is located. For several decades prior to that, temperatures throughout the western Arctic were falling.[5] They have been steady or falling since then in other regions, such as Greenland.

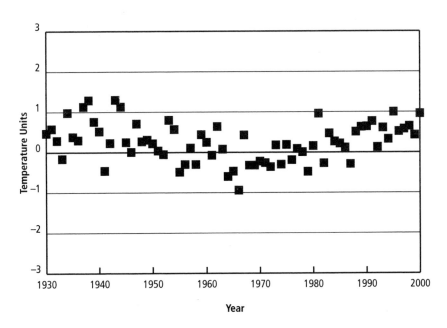

Figure 2.5. Average Arctic air temperature anomalies from 1930 to 2000, as computed by the Hadley Centre in the UK.

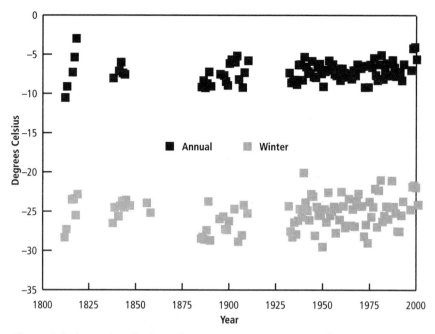

Figure 2.6. Annual and wintertime average temperatures for Fort Churchill (1800 to 1883) and Churchill (1884 to 2000), Manitoba. Data from NASA's Goddard Institute of Space Science.

We showed the annual average temperature series for the entire Arctic since 1970 in Figure 2.3. Figure 2.5 shows the same data all the way back to 1930. The big increase in the average took place from 1900 to 1940, and since then there has been a bit of cycling up and down, though the overall trend since then is downward. The years with the highest averages over the past century were between 1935 and 1945, which would predate the clear memory of many elders. Presumably the Inuit lifestyle (and the polar bear population) survived that warm interval, and will in all likelihood survive the present one as well, if temperature is the only issue.

How about those polar bears? The temperature records for Churchill and adjacent Fort Churchill go back to the late 1700s. Winter and annual averages for the past 200 years are shown in Figure 2.6. The record is intermittent, but most years are available. The warmest year was 1818. The average drifts around between –5°C and –10°C. The years 1998 and 1999 were a bit warm, but temperatures since then have dropped. March, April and May of 2002 were the coldest, on average, since 1967. Twentieth-century winter temperatures peaked in 1940 at –20°C and otherwise have remained around –25°C, except during the temporary El Niño event of 1998. There is not much danger of Churchill ceasing to have winter anytime soon.

Now you know everything you need to know about temperatures at Churchill, Manitoba, to decide if there is a crisis of unusual warming up there. To figure out what (if anything) is happening to the ice, you need to get some data on wind directions and ocean currents in the southwest region of Hudson Bay over the past few decades, since this is what drives variations in springtime ice thickness. These data are probably available somewhere, and will likely show that we are on the low side of a multi-decade cycle. As for polar bears, understanding how they are making out requires data on things like natural population cycles, predator-prey cycles, hunting rates and so forth.

This example shows that self-reinforcing currents of certainty can get set up between Official Science and political leaders even when the facts are very uncertain. Energized by the media and activists, a convection cycle can take some very ambiguous information—in this case the suggestion of unusual warming in the Arctic generally and around Churchill in particular—and quickly spin it up into a thick cloud of Doctrinaire certainty that rains out regular science and thunders "aggressive action" upon emitters of carbon dioxide.

To a regular scientist, the tempest over Churchill is calmed by looking at the data. If something unusual is happening, it is not driven by a rise in air temperatures. If there is a retreat in the extent of springtime ice, perhaps there are oceanic causes. We need to gather data, observe, test hypotheses, and study it carefully. That is the way searching for truth always proceeds. There are always statements and corrections.

Perhaps something odd is going on. Perhaps not. Similarly, if the Arctic as a whole is undergoing a warming phase, it seems to be a repeat of what happened earlier in the century, and even still we haven't recovered to temperature levels experienced in the first half of the last century. But we can investigate all this. What drives temperature fluctuations in the Arctic? What is the role of the brightening of the sun over the 20th century? Or the Arctic Oscillation, or the Pacific Decadal Oscillation, or the El Niño-Southern Oscillation? These are questions we can and should ask (see, for instance, Chapter 6). They are regular science questions. They are not the stuff of the nightly news. They will not serve the purposes of green groups that depend on activism to raise funds. They might frustrate politicians who want reasons to implement a global solution. They are irrelevant details to Official Science. But they are the sorts of questions that need to be answered if sound policy is to be made. The real test for the openness of people's minds is whether contrary evidence can reverse views. The supporters of the warming idea are being tested.

Our reconnaissance flight through the wild winds of the global warming storm is over. The elements of the system, their interactions and self-reinforcing motions have been mapped. It is now time to look past the storm and re-examine the issue that gave rise to it in the first place. The remainder of this book is the second, larger reconnaissance mission: mapping out what we know and don't know about climate change science and the appropriate policy responses; and suggesting how to renew a healthy scientific process for studying both.

---

[1] http://news.bbc.co.uk/hi/english/sci/tech/newsid_1130000/ 1130501.stm.

[2] Houghton, John. *Global Warming: The Complete Briefing*. Cambridge University Press, 1997, 158–59.

[3] http://www.environmentaldefense.org/pubs/AnnualReport/ 1991/c_air.html.

[4] National Centre for Environmental Assessment (NCEA). NCEA/ Global Research Solicitation—NCEA-01-01, July 3, 2001.

[5] Kahl, J.D., D.J. Charlevoix, N.A.Zaitseva, R.C. Schnell and M.C. Serreze. "Absence of Evidence for Greenhouse Warming over the Arctic Ocean in the Past 40 Years." *Nature* 361, 28 January 1993, 335–37.

3 Climate Theory Versus Models and Metaphors

"One thing I have learned in a long life: that all our science, measured against reality, is primitive and childlike—and yet it is the most precious thing we have."

—Albert Einstein

Climate Theory and Climate Models

Front-page headlines around the world in January 2002 proclaimed the apparently amazing news that it is getting colder in Antarctica. The papers told us that this is a big surprise because computerized climate models had predicted it would be warming there. But what was so surprising about it was that for many people this news was surprising. After all, the previous week, some computer model predicted snow that never came for Toronto. *That* didn't make headlines around the world. No one expects computer models of the weather to be that certain. Yet many have come to expect climate models, which treat a far more difficult scientific problem, to be so certain that a gap between predictions and reality over a small region of the planet is a worldwide news event.

The truth is, we have much less reason to ascribe certainty to climate models than we do to weather models. So why the headlines? How did this unwarranted belief in the certainty of climate models arise? Among other things, this is the Doctrine at work. If, as the story goes, scientists fully understand how the climate works, and they all agree on how to model it, and the models all tell us Antarctica is warming up, then, by golly, Antarctica should be warming. Those same news items were fascinating for the comments by some scientists to the effect that Antarctica really *is* warming, but for the masking effects of some sort of...cooling.

We will leave this sort of curious reasoning to others. Instead, this chapter will go straight to the basic and essential issues that have been widely overlooked. Be forewarned that they may be rather hard work, but you'll be all right; it won't hurt. We need to draw some important distinctions between three things: **theories, models** and **metaphors**. If any of these three deserves to be called certain, it is the first—and science has taught us to be very tentative in such a use of this word. In this sense, models have a lower standing in science, and metaphors lower still. They are *useful*, but can easily be misleading if they are not used wisely. Ultimately, in the physical sciences, it is only theory that is relentlessly tested by controlled experiments. In contrast, models are constructed to fit only some observations and metaphors need have little to do with experiment. A theory defines what a proper use of a model is and the truthfulness or otherwise of a metaphor.

We will start with the question of whether there is such a thing as a theory of climate. The answer is brief (*no*), but explaining why will require your sustained attention over many pages. We warned you before that there are no shortcuts in this business. In fact we will explain how attempts at shortcuts have addled the wits of the collective consciousness on this subject. Misplaced confidence in models, as substitutes for theory, and the overuse of misleading metaphors, not to mention a lot of lazy deference, to Official Science, has done damage to climate science and the policymaking process. Here is where we begin undoing the damage.

In talking about theory and models, we need to use some ideas from mathematics. There is no choice; mathematics is the language true theories are written in. We have to talk about the theory in order to explain why there isn't one for climate, and so we must talk about mathematical things too. Now, mathematics is not the evil and ancient language of Mordor, although people squirm like patients in a dentist's chair when you suggest that it would be fun to use, so we will not actually use it, although we will have to talk about aspects of it to make progress. Herein lies a version of the scientist's dilemma discussed in Chapter 2. We will begin with a few ideas that nearly everyone is familiar with and sketch out some additional thoughts. Chances are they will be enough.

Something equalling something else is part of the general knowledge. We say that two plus two equals four, and everyone understands that is what an equation is all about. A more mature way of thinking has symbols standing in for the numbers. Nearly everyone has seen Einstein's famous equation $E = mc^2$. To describe a physical system often requires a group of interrelated equations. We say that we can "solve" the group of equations if we can rearrange them so that what we want to know is revealed to be equal to some complicated arrangement of things we already know. Some groups of equations have no solutions, some have an infinity of solutions and some have only one. It would appear to be simpler if every set of equations had only one solution; alas, it is not so. And, as is often the case, these additional qualities tell us things.

In the natural world, we are interested not just in the values of some symbols, which we will call **variables**, that we want to know, we are also interested in how the variables change over time, space and physical conditions. We are interested in their dynamics. The result is that we are no longer concerned about simple relationships between variables. We are interested in a wider class of relationships called differential equations, which involve the tools of calculus. The theories of basic science are written in differential equations. In principle, everything about the physical circumstances of our world can be represented in terms of differential equations. We can describe the motions of the planets and the flow of the blood in your veins with them.

A solution of a differential equation is not a number or a few numbers; it is a **function**. A function is a rule between variables. For example, if you have a rule between position and time that tells you how to find position at every time, we say that the rule gives the position as a function of time. Solving a differential equation amounts to finding such functions. As you can see in this example, we can predict the future if our equation is right and we can solve the equation. That's why we want a theory with correct equations. Forecasting the future then means nothing more than seeing where you are now and using the differential equation to tell you about the future through its solutions.

It all sounds very nice and straightforward. We will see below why it is not.

Marooned Halfway up Mount Climate Theory

We cannot see atoms and molecules or their dynamics directly, although our lives and senses are made rich by their effects: everything from colour to temperature, texture, taste and smell. The movements of the atomic world that may be experienced in the air around us are on length scales of 10^{-6} centimetres and time scales of less than 10^{-8} seconds. On the other hand, we experience daily life on times of the order of a pulse of the human heart, one second, and the width of a hand, 10 centimetres. Climate change is as remote from our experience as the world of atomic movements, and we are just as unable to see or experience it directly in our daily lives. But that is because climate is too large and slow to see, rather than too small and quick. Instead of 10 million times smaller than a hand, we think about length scales 10 million times larger, and instead of 100 million times shorter in time, we will think in terms 100 million times longer. So the world of climate begins on length scales in excess of 10^8 centimetres, which is a bit larger than the horizontal resolution of a good general circulation climate model, and 10^8 seconds, which represents a time scale of a few years.

When you look out a window, the weather you see is not climate. As with atoms and molecules, you can only get some idea of it through indirect means. There may be palm trees or there may be snow outside to give you a clue, but you cannot actually see climate itself with your own eyes. Our knowledge and experience of it is fundamentally indirect, accumulated from years of experience or from the prevailing plant life. We often defer to elders and look at records accumulated over generations to get a sense of it.

Just as we can imagine riding on molecules, we can envisage what it must be like living on scales sufficient to actually see climate working. Think of your hand as becoming as large as a small country, and your heart beating by the year rather than the second. The last ice age would have been "hours" ago and the dinosaurs would have been alive a couple of "years" back. Of course, you would not have been able to see the dinosaurs, nor would you be able to make out humans or most of their works. They would be to you as bacteria or viruses.

The structure you perceive would be what no one has ever experienced before. We can only guess at what it would be like. It is a hidden world on the grand scale, little known in comparison to the small hidden world of atomic movements. For the small hidden world, we have very great confidence about the rules—what the components of that world of atomic movements are and what governs their behaviours. However, in that large hidden world, we have the barest hints at structures and only the slightest clue about what the rules would be like.

Human lives are lived in a middle realm between these two hidden worlds. We know the structures and theory for the small hidden world so very well that we can boldly attempt to construct our middle realm from it. We use averaging to do this, since we don't directly experience atomic motions, but collective behaviours instead. Averaging is one of those mental steps we do all the time to make large lists of numbers manageable. From an array of numbers, we extract a single central quantity that stands in for the list. We could not, for example, keep track of the specific daily prices of bread, milk, shoes, gasoline, beer and the other things we buy regularly. But we can keep some average prices in mind, and that makes our household budgets manageable.

We do not need to pay attention to the details of how we construct mental averages for things like the price of beer, since there is no real harm in being only approximately right. The requirements are higher when doing averaging in scientific fields, however. We average over representations of theoretical variables for the atomic world and thereby make approximations, suppressing things that would not show up significantly. We cannot do this any old way, but we have a tremendous advantage in choosing the methods because we know what the answer has to be. It should match what we experience in the middle realm.

It would be a mistake to underestimate the importance of this. The route upward toward our middle realm contains dead ends and false trails. If you are not careful and the averaging isn't done correctly, the world of our experience gets averaged away into a gray amorphous nothing, or into things that don't converge to anything at all, making it seem that there is no structure at all in our middle realm. It is precisely our experience otherwise that instructs us on how the journey must be taken.

Averaging in the wrong way over the wrong things have produced results that are inconsistent with what most people know about fluids such as air and water. They are governed by nonlinear as opposed to linear equations. These represent major distinctions in types of great importance. However, the misleading averages can yield exact averaged equations of quantities describing fluids that are linear! This is especially remarkable because the correct differential equations for fluids are some of the most notorious examples of nonlinear differential equations there are.

In an unmathematical world, a differential equation being "notorious" seems hard to imagine, but some really are. You won't find them in *People* magazine, but nonlinear equations have notoriety among those who know about them, because we can't solve most of them. We are left to rely on computer approximations of solutions. Furthermore, unlike linear equations, they can and do exhibit a kind of peculiar unpredictability in their solutions, not unlike randomness, known as **chaos**. We will say more about that a bit later.

It would be so very much easier if we could work with linear equations, but it is not to be. The averages that are used in these linear equations turn out to be quantities that our instruments in the middle world don't actually measure, and the physically meaningful structure we know and experience from fluid behaviour is not represented by them. Furthermore, these summed-up quantities cannot stand up on their own. There is no extra rule that relates the averaged variables to each other. We cannot solve for them because we end up with more new averaged variables than equations. The only way to say what the averages will do is to start over again from first principles. So the averaged equations turn out to be useless for the very purpose they were originally intended.

They don't tell us anything new about how things are related in the middle realm because we always need to go back to the scales of atomic movements to find out what we wanted to know. This means that those particular averages have no physical significance. They cannot then be considered to be physically worthwhile variables. They are just statistics and do not pertain to the underlying physics except through some unenlightening definition. It is not worth the trouble computing them in the first place.

We say that the averaged equations don't close on themselves because we have to go outside of the circle of averaged variables to compute things. The equations are said to have a "closure" problem. The linear equations are little more than curiosities for theorists. However, the lesson for us is about averages themselves: There are many different types of them, and not just any averaging scheme will give a physically meaningful result.

The nonlinear fluid mechanical equations that come out of the averaging process and that do agree with experiments in the middle realm are known as the Navier-Stokes equations. The experiments helped to teach us how to connect them to the world of atomic movements reasonably well. The theory that connects the two realms, the small hidden one and the middle one, is called *kinetic theory*, from the Greek word for "movement."

Kinetic theory marks out the path from the base camp—our knowledge of the world of atomic motions—to the mid-level camp: our knowledge of the middle realm. But the Navier-Stokes equations were already known before kinetic theory. It may seem to some that this diminishes the accomplishment of connecting those two worlds. However, kinetic theory teaches us what it takes, theoretically, to move between different regimes of nature. We needed to talk about kinetic theory here because it gives us a sense of what it must take to move between the middle realm and the large hidden world of climate. Clearly, that is something similar to the problem of kinetic theory. We know we are going to do averaging, and kinetic theory acts as a warning for some of pitfalls in that approach. But there are many differences too, and nearly all of them make the climate problem more difficult than the problem of kinetic theory.

The most obvious difference is that we have no guide in the larger climate world to any key structures and relationships. There is no one living on climate scales to observe structures, do experiments or establish physically meaningful structure for us. Without a climate structure analogous to Navier-Stokes to act as a beacon to climb toward in our averaging schemes, we are little better than bacteria in a test tube trying to deduce from first principles what the laboratory ought to be like.

Average Performance

Still, we have to do something. So in climate work, everything is being averaged in every conceivable form, and some inconceivable ones too—self-consistently and otherwise. The trouble with averages is they never tell you when you are using the right or the wrong one. The tragedy of averages is that many think there is no wrong one. The example from fluid dynamics shows us, however, that there are wrong ones. To find one that is right, you need some other information, in addition to the averaging itself. The unwritten hope has been that in the absence of this added information, somehow the right averaging schemes across data and theory from the middle realm will emerge through good fortune anyway. We will know it by some kind of dovetailing in the results that gives us some kind of as-yet-unforeseen vision of the large hidden world of climate, of which hitherto we have had only the barest glimpses. This would then, we hope, lead to a theory.

But for now we have no theory for climate comparable to what Navier-Stokes theory is for middle realm fluids. Unlike kinetic theory, where our journey through averages is toward a known theory, for climate the journey is away from a known theory into the unknown.

There is another, still more troubling difference between the climb to the large hidden scales and kinetic theory. The kinetic equations for movements of individual atoms can be solved. This puts them into a form that makes kinetic theory useable. But we do not know how to solve the base equations (Navier-Stokes) for our climb to the larger scales.

What does that mean? As mentioned above, solutions for differential equations are not numbers but functions, which are nothing more than rules that say how variables are related to each other. Every scientist or engineer learning about differential equations in the classroom is treated to a multi-year parade of differential equations, all having solutions that can be written down after applying simple and intermediate mathematical methods.

Only a few go on to advanced courses where they encounter the much more numerous class of differential equations where we do not know how to find any rule to write down. For some of these, we can show that there is a rule, even though we cannot figure out how to get one to write down. On the extreme, nasty end of these equations, we cannot be certain that a rule even exists. The Navier-Stokes equations belong there, so nobody knows if there are actually solutions of the Navier-Stokes equations in all circumstances of physical interest, despite their having been around for a century and a half.

Many people in engineering and meteorology use some version of Navier-Stokes all the time, either in highly specialized cases where they can extract solutions or approximate solutions by various means, or by using computers. They often use simplified versions that work better on computers. So, generally, they don't care whether formal solutions exist or not. The idea of establishing the "existence" of solutions is typically regarded as a pastime of obsessive mathematicians. Engineers and meteorologists can always say the self-evident presence of fluid flow itself is enough to set aside any concerns of this type.

They have a point. The experimental evidence from the middle realm is a powerful argument to set aside these apparently subtle concerns.

Navier-Stokes theory was born in the middle realm and it agrees with many controlled experiments in that domain, although certainly not all. The fine points of underlying atomic movements or the existence of solutions can, in that framework, seem to be a kind of exotic embroidery that is not important to routine problems.

But this is the heart of the matter. Climate research is anything but a routine application of classical theories like fluid mechanics, even though some may be tempted to think it is. It has to be regarded in the "exotic" category of scientific problems, in part because we are trying to look for scientifically meaningful structure that no one can see or has ever seen, and that may not even exist. Everyone agrees that some kind of averaging may help, but this idea is in itself not very helpful, even though unchaperoned averaging has been going on in the back allies for decades. And, in extrapolating out of the middle realm, the experimental-evidence argument for Navier-Stokes theory is not quite so convincing.

Furthermore, experiment and theory have been struggling since the 19th century, literally for generations, with a complicated behaviour of fluids called **turbulence**. When a fluid is turbulent (nearly all natural fluids are), not only are we unable to provide solutions of Navier-Stokes to confirm the behaviour theoretically, but we are also unable to experimentally measure the conditions in the fluid in such a way that we can fully capture what is going on. The experiments are bedevilled by the fact that a turbulent fluid is active on scales smaller than the size of the finest experimental probes. Thus the measurements themselves are not of the actual variables but of some kind of unspecified, instrument-dependent average of the variables, in only one small region of the fluid.

It was early in the game when people like Osborne Reynolds in the 19th century thought it might not be the best idea to try to compute solutions of the Navier-Stokes equations in a turbulent medium, but to look instead at an averaged version of them. From a kinetic theory point of view, it is as if the first round of averaging up from atomic scales didn't work—so people tried a redo. The immediate mathematical problem is that after the redo averaging, there are more averaged variables than there are equations. This persists in various clever expansions of the equations too. There are always more variables than equations. So the new system can't be solved on its own terms. You have to go back to where you started to make it work. In other words, on its own, it leads nowhere.

This is the famous "closure" problem of turbulence theory. Over the past century, a lot of very smart people have spent their careers unsuccessfully trying to find a general way around this problem. As we have already encountered a closure problem above in connection with kinetic theory, this seems as much a broader pitfall of the averaging concept as a special difficulty of the turbulence problem.

If you have an experimental setup for what you are studying, the closure problem can be solved in practice. Empirical relationships can be introduced by measurements of turbulent flows to complete the problem, making practical engineering projects and computational fluid dynamics possible in turbulent cases. However, even for the simplest cases of turbulent flow (such as measuring the simple flow of water in a kitchen pipe), the fact remains that we cannot predict the flow from the basic physics alone; moreover, we cannot even predict the average turbulent flow from the basic physics! We are in a sorry state here compared to the other physical sciences. It reflects the enormity of the problem.

In this regard, it is crucial to bear in mind that there is no experimental setup for global climate, so all we really have are those first principles. You can take all the measurements you want today—fill terabytes of disk space if you want—but that does not serve as such an experimental apparatus. Engineering apparatuses can be controlled, and those running them make measurements of known variables over a range of controlled, physically relevant conditions. In contrast, we have only today's climate to sample directly, provided we are clever enough to even know how to average middle-realm data in a physically meaningful way to represent climate.

Global climate is not treatable by conventional means.

Kolmogorov's Trashcan and What Was In It

Do we have any clues at all on how to start the climb toward the summit of Mount Climate Theory? For a while in the 20th century, it was looking good. Computers were appearing on the scene, and data were more systematically collected. Many scientists believed that putting in ever more copious detail might pull off the climb. Sure, there would always be something missing, but with the aid of more data and the growing computational power, perhaps it wouldn't matter. It didn't before. What ultimately did happen in science surprised everyone, and it all had to do with turbulence.

It is extraordinary that theoreticians avoid what is regarded as a complete theory to make progress, although it is entirely normal in this field, again reflecting the enormity of the scientific problem. Remarkably, it took setting aside the Navier-Stokes equations to generate two of the most significant theoretical insights of the 20th century into turbulent flow.

The first of these was the Kolmogorov theory from the 1940s. In fact it is often referred to as Kolmogorov '41, like a type of wine. Kolmogorov and those that followed used what is called dimensional analysis and a physical way of characterizing turbulence as turning whorls of fluids called vortices or eddies. The idea is that turbulence is basically vortices within vortices and so on, down and down, smaller and smaller, until the movements became dominated by internal frictions that stop the very tiniest whorls.

(Incidentally, it is in such vortices that Navier-Stokes solutions have been proposed to fail. The idea, from the 1930s, was that at each vortex center is a point where some component of the velocity field becomes infinite. This idea was influential in advancing other mathematics, but it has not been proven or disproven for the Navier-Stokes equations to this day. However, very recently, enormous accelerations, taking place only over minute scales, have been observed in turbulent water. They include accelerations of 1,500 times the acceleration of gravity. Fighter plane pilots deal with not much more than a few times the acceleration of gravity. This suggests something dramatic does indeed happen on very small scales at some places within turbulent flows. But since we can't solve the equations, formally, we do not really know for sure.)

The Kolmogorov theory produced simple relationships that characterized how an approximate number of whorls varied with a property of the fluid flow conditions called the Reynolds number. The relationship held only for so-called fully developed turbulence. The theory also established the idea of a nonlinear link, through Navier-Stokes, between the big whorls and smaller ones. The big ones have their big movements transformed into smaller movements and the smaller ones into still smaller and so on. The energy moves from big scales to smaller and smaller ones. This flow of energy from larger to smaller scales is known as an energy cascade. In this way, movements that are far too big to be affected by the internal frictions of fluids eventually die out anyway by making themselves over into smaller and smaller whorls that are eventually dissipated.

This produced a powerful mental picture for scientists, which the

Kolmogorov theory did not actually imply and which Kolmogorov himself probably didn't intend. The idea of an energy cascade seemed to take on a direction, in many minds, in the sense of large to small. The idea of an energy cascade was easy to confuse with a "cause" cascade. In the minds of some, not only does energy move down to smaller and smaller scales but so does the cause-effect chain. Big things only cause smaller things, smaller and smaller until dissipation, while small things never cause bigger effects on upward. There was only one direction in that picture.

Such a picture, if it were true, would have made life so much easier than it actually turned out to be in studying turbulence. For example, computer modellers of turbulent fluid flow could just cut off a calculation of Navier-Stokes at the small scale of choice and drain out the energy in a clever way at the cutoff, and things should work out just fine. Below that small scale of choice would be a kind of trashcan of unimportant small movements.

This brings us to the second promised theoretical insight about turbulence. While it was presented in the early 1960s, it was not really until the 1980s that its effects began to sink in, and in some quarters, the ideas have not yet arrived. A meteorologist named Edward Lorenz experimented with collections of differential equations called systems. These equations were a mathematician's toy model of atmospheric turbulence. In particular they were meant to characterize convection, which in the atmosphere is a special kind of turbulent motion normally viewed as a crucial mechanism for moving heat away from the earth's surface.

The equations are (intentionally) much simpler in many ways than Navier-Stokes. For example, the Lorenz equations have solutions that exist only in a three-dimensional mathematical space, while the Navier-Stokes solutions exist in one with infinite dimensions. Unlike Navier-Stokes, it is straightforward to show that there are unique solutions for the Lorenz equations. But like Navier-Stokes, we don't know how to write them down, even if we can use a computer to estimate them very accurately.

The computed solutions of the Lorenz equations revealed a remarkable thing. Small things had big consequences after all. It was the opposite of the Kolmogorov trashcan picture, although it didn't contradict the Kolmogorov theory itself. It only spoke to the rarely voiced but widely held misconception of the one-way cause and effect relationship, where little things are supposed always to be caused by bigger things.

We always knew this is not true. It is, for example, clear in various versions of an old verse probably originating from Benjamin Franklin:

> For want of a nail, the shoe was lost;
> For want of the shoe, the horse was lost;
> For want of the horse, the rider was lost;
> For want of the rider, the battle was lost;
> For want of the battle, the kingdom was lost,
> All for the want of a nail.

A specific chain of events, ending in a significant result, can be traceable to the smallest origins. It doesn't mean that small things have to be powerful in themselves. The amount of energy is beside the point. It is the particular sequence of events an initial event sets in motion that can give even the smallest event its significance. Pushing over the first domino in a long row of dominoes is a classic example; turning a key to start powerful machinery is another. They are not just started but controlled in this manner. This is not new; it is part of life; it is everywhere.

In mathematics, differential equations can be viewed as rules about movement between events, leading to a sequence of events not unlike the sequence of lines in the verse. Viewed in this sense, differential equations are referred to as dynamical systems. The rule of movement is applied to some starting point leading to a new event, which leads to the next, and so on. The resulting sequence is the form a solution takes for dynamical systems.

Of course the nail-and-shoe verse is really about *two* sequences of events, one without the nail, and one with the nail wherein the kingdom is *not* lost. The latter is only present implicitly, but it is essential for the verse's meaning. The verse announces the difference in each line with, "For want of." So this old verse not only tells us about comparing sequences of events but shows that the differences between them form a sequence as well. Subtracting the nail begins a sequence of five differences ending with subtracting the kingdom. We say that whether the kingdom is kept or not is "sensitive to the initial conditions" of whether the nail is present or not.

After the five lines, the differences end. The verse doesn't go on to talk about an empire being lost, or a planet or a solar system. It stops at the kingdom being lost. The magnitude of the consequences of sensitivity to initial conditions is limited. That is how sensitivity to initial conditions happens in the real world: Eventually you run out of dominoes; turning the key in the ignition just starts the car, it doesn't launch you into space.

Differential equations that are sensitive to initial conditions (for which we can write down solutions) have been known since the beginning. But the differences between solutions of those previously known equations are not limited. The differences just keep growing. This is a crucial distinction because the Lorenz equations are about the turbulence in connection with weather. We are only talking about differences between sun or rain tomorrow. We are not talking about rain or supernova tomorrow. Sensitivity to initial conditions would not make sense for weather if the differences kept right on growing. The solutions of the Lorenz equations therefore are sensitive to initial conditions *and* the differences are limited.

This was a new result, which made this type of sensitivity possible to envision for weather, not to mention turbulence. Moreover, sensitivity to initial conditions is *typical* for the solutions of the Lorenz equations. This is different than the cases of the nail, the dominoes or the car ignition. They were either just flukes or contrived arrangements. In contrast, the Lorenz picture is pervasive and it is not a special setup—nor is the weather.

This pervasive sensitivity to initial conditions, with differences that are limited, had not been fully appreciated before Lorenz's work. Lorenz at first pointed out that weather forecasts, no matter how carefully prepared from some initial condition, could be thrown off with the flap of a seagull's wing. Later the metaphor was superceded by the more graceful image of a butterfly's wing and the phenomenon came to be called "the butterfly effect." With the help of many other parallel lines of research from others in other fields of science, this effect, subsequently found in many other dynamical systems, came to be known as chaos.

It was the end of simple predictability in much of classical science. By the 1980s, scientists were finding chaos everywhere, even in such icons of predictability as the pendulum. It was a revolution and a renaissance for classical physics that ran through the last two decades of the 20th century and continues today. It spread through chemistry, biology and onward. And a key part of this revolution came from the struggle to understand turbulence in the atmosphere.

Although the Lorenz equations were only a toy model of Navier-Stokes, it was understood that they presented phenomena that would not likely be more benign and simplistic in real turbulence. People found many similar and more complicated phenomena in the real thing. However, the idea of small affecting large spread into many other unlikely fields having nothing to do with fluid dynamics as similar previously overlooked things were found. It was nothing short of a revolution in scientific thinking. This was elucidated by one of the greatest fluid dynamicists of the 20th century, Sir James Lighthill, who made a remarkable speech in 1986 at the Royal Society about it. Addressing himself in part to the "public at large," he offered an unprecedented apology, on behalf of classical physicists, for over-stating our ability to make predictions about natural systems:

> We are deeply conscious today that the enthusiasm of our fore-bears for the marvelous achievements of Newtonian mechanics led them to generalizations in this area of predictability which, indeed, we may have generally tended to believe before 1960, but which we now recognize as false. We collectively wish to apologize for having misled the general educated public by spreading ideas about the determinism of systems satisfying Newton's laws that, after 1960, were proved incorrect. In this lecture, I am trying to make belated amends by explaining both the very different picture that we now discern, and the reasons for it being uncovered so late.[1]

Theoretical treatments of turbulence were attempted from classical science. The outcome of this was wonderfully ironic. Instead of classical physics cracking the turbulence problem, the turbulence problem cracked classical physics. To the delight of scientists everywhere, turbulence bit back.

This meant that the Kolmogorov picture does not have a direction after all. Small things routinely led to big things, as well as the reverse. Change ran both ways in terms of scales. There was no Kolomogorov trashcan: There was no level of detail that can be safely ignored. Ideally, fluids ought to be treated on all scales. Some had hoped that we could do otherwise without consequence. They dreamt that the individual move-ments in the small hidden world would cancel themselves out as thermal noise in the middle realm, and maybe averaging would bury them so that they would not complicate things. But it was precisely atomic scale frictions that cause things like fluid viscosity, which ultimately grew to make the vortices that constitute turbulence.

There is a Greek myth about a great gift that could not be used. The Navier-Stokes equations have this mythic quality. A great gift—the fundamental equations of fluid dynamics—was given to us in the 19th century, with the catch that no one can figure out how it can be fully used. That wonderful gift remains tantalizingly beyond us to this day. Only Cassandra could fully appreciate the irony. That irony, and the revolution in classical science wherein small things were found to have big effects, are more than esoteric curiosities for those looking for a theory of climate. And, even now, the old ways persist in some areas of climate research. The revolution has not yet run its course.

Fairy Tales of Computation and the Devil's Ball of Yarn

The Enchanted Computer

We are marooned halfway up the mountain. We have no theory for the big hidden world of climate. Moreover, Navier-Stokes isn't even a complete theory for the middle realm. Obviously, much more goes on in the middle realm than can be represented by fluid dynamics generally, and Navier-Stokes in particular. However, the cream of 19th century and early 20th century theories of classical science is available to us to (nearly) complete the picture.

In principle, we can understand and forecast what happens in the middle realm. That is, we have the theoretical tools to look at each part: fluid dynamics, solid mechanics, thermodynamics, radiative transfer, classical chemistry, classical electrodynamics, geophysics and celestial mechanics. So why don't we just shovel them all together into a great big giant computer to tell us everything there is to know about the physical world? With all the details (all of them), this would be a direct computation of the theories and not simply a model. It would not be modelling; the computer's solutions would behave exactly as the real world does, in principle.

There was at one time a great dream to do just that—put all of these great human accomplishments to work representing the middle realm. In that dream, they would be placed into an ideal computational machine of astronomical proportions. However, it turns out that we cannot construct such a machine, and no one expects us to be able to anytime soon. But we can dream.

Maybe a fairy godmother will wave her magic wand to turn something from the real world, say a pumpkin (or an Apple), into an Enchanted Computing Machine (ECM) that can magically cope with all the details needed to compute all of the theory while securing all the necessary initial data to implement it. We would in effect convert the entire middle realm into information and dynamics represented within the machine's components. Then the monster calculations would begin. We would demand from the machine the answer to the question of climate and everything, not unlike what was demanded of Douglas Adams's computer character, Deep Thought, in *A Hitchhiker's Guide to the Galaxy*.

With the ECM, people could do genuine experiments on the real global climate, at least in principle. There is, after all, no other way to do them. We don't live on a grand enough scale to carry out climate experiments. We don't have any spare Earths to carry them out on either. The inability to carry out experiments is one of the defining characteristics of the type of scientific problem that climate represents.

It is one of the things that sets it apart from other problems of science. Having no climate experiments doesn't mean that we cannot gather observations about it. We do gather it, in astronomical quantities. However these observations are not standins for experiments. Experiments are more than an exercise in free-range data gathering. You need some idea of what data to collect and what you are trying to do with them. The setup needs to be controlled to allow the measurements to focus in on only those aspects of the problem you are trying to experiment with. If other things get into it, you soon have no idea of what is really going on.

For climate, there is no setup at all. Everything is wide open. Climate observations are not controlled experiments—and treating them as if they are has caused plenty of confusion. It does not reflect badly on those people who gather climate data that they are not doing experimentation. Instead, it speaks to the fundamental difficulty of the climate problem. We cannot just shut down and divert parts of the climate to test things. Imagine receiving a circular asking us to make other arrangements, as the Global Climate Experiment Commission will be shutting down the Gulf Stream for five years to see what happens to averaged upper-level winds, and it apologizes for any inconvenience.

It's not going to happen.

In principle, the ECM would eliminate this basic problem by providing a way to test things. Even if we only have an understanding of each of the classical theories in isolation, having them work together in this way would give us a chance to grasp a complete picture of what is going on in the middle realm and do the experiments we can do in no other way.

However, even if a fairy godmother gave us an ECM, it would only be a tool for studying the middle realm. We would still be unable to perceive climate directly. To go further, to go toward climate itself, is to pursue a dream within the ECM dream. We would have everything at our disposal to average and average until, with luck, we could come up with an averaging scheme that induces new, as yet undiscovered variables, with relationships only to each other, that reflect the essential nature of climate. That would indeed be the ultimate theory for climate. But that is just the dream within the dream. We have no theory for climate. We have no ECM.

Yet people do still pine for one. The ECM dream comes in two forms: strong and weak. The dream of the strong ECM, wherein perfect predictability is achieved through computation in complete detail, died with the chaos revolution. However, not unlike Elvis sightings, you can still hear rumours that someone has created one, or that somewhere they are close to creating one.

People get quite enamored with the computer models we do have. Even worse, some who should know better think naively that we have already had an ECM for a long time. Perhaps they are confusing it with a "GCM," which, as we will explain shortly, is quite a different thing. But finding an ECM is not going to happen. Infinite-sized computers are nonstarters, and the near ideal of finite-sized computers large enough to approximate them was killed by the sensitivity to initial conditions.

The latter is a difficult concept for people to understand the significance of, especially these days, when children learn arithmetic on calculators. Unfortunately they usually do not learn that calculators and computers can and do give wrong answers because of their finite size. Because of that, and the fact that we have a finite number of probes, there will always be something, maybe very small, that has to be left out when implementing the mathematically based classical theories. And when small things have large consequences, this is not a good prospect.

Nonetheless, you can still hear the echoes of the strong ECM dream in the remarks of some scientists, even some who may not even believe in the ECM idea themselves. Accurate prediction, they say, is just a matter of having a big enough machine and enough data. That is the discredited idea. It will not be enough, even if we have so much data we will have to learn to assimilate observations "like drinking water from a fire hose," as another of the pet expressions goes.

There is also is a weak ECM dream, which is more sophisticated in that it reflects an awareness of sensitivity to initial conditions. The idea is that the ECM should compute everything (in principle) as before, but, to take account of the butterfly that fluttered by some kid in Kamloops, you do the computation over again from a slightly different starting point. And then you do it again from yet another, again and again, until you have all of the likely starting points covered. Then you take an average over all of the resulting outcomes. You then say you have computed the averaged middle realm, *in principle*. Hopefully, so the dream goes, the averaged middle realm will be insensitive to the small things. Think about doing this to the verse, with the nail and again without it, then conclude, on average, you have half a kingdom. It could work, in a sense.

It is a clever idea, but the weak ECM will not work in principle. Any ECM is truly a myth. Mathematics is too large to fit into a computer of finite size, and infinitely large computers will likely never be on the drawing boards.

But do we really need to know every detail? Do we really need to know all of the mathematical minutiae? Do we really need to know where every atom is and what it is doing? Before the chaos revolution, we would ask whether something was big enough to matter. Now we must also ask if something is small enough *not* to matter.

How small is small enough? This turns out to be a very tough question to answer, and we are only getting started at it. We need to ask the question not of individual theories, but of all the theories fully integrated together, as the answers can be very different for each theory and together they play off of each other. However, we don't actually know how to do a grand treatment of all of the classical theories together in practice, let alone install them together in an ECM, even if our fairy godmother offered us one.

These theories do not just snap together like Lego. They are tough to handle in an uncontrolled, non-experimental environment, but together they become breathtakingly complex. It is typical of the deep nature of the problem that global effects from their interactions need to be developed from their natural connection on the smallest, even molecular scales. Having to deal with such a range of scales simultaneously is what makes turbulence so intractable. It is only compounded with all theories in play.

The Devil's Ball of Yarn

For example, while we understand very well the thermodynamics of changes of state among solids, liquids and gases in controlled laboratory conditions, stitching this type of basic thermodynamical theory together with fluid dynamics leads to profound complexity.

There is a field of fluid dynamics called *turbulent diffusion*, wherein scientists study the movements of contaminants or pollutants pulled along by a turbulent fluid. A remarkable thing discovered in recent years is that the polluted air doesn't immediately become mixed with the plain air in turbulent motion. Instead it gets pulled and stretched out like toffee, longer and longer as it is turned back on itself again and again as the whorls and vortices do their work. The result is groupings of long thin filaments composed of still thinner long filaments. Within a small filament the polluted air is as it was, and outside of it, the plain air is also untouched. Molecular diffusion will eventually eliminate the differences, but turbulent movements are much quicker.

What does this have to do with things other than pollution? For simplicity, suppose turbulent air comes in two basic forms: moist and dry. The atmosphere stores and releases large amounts of energy using water vapour. When water is vapourized, it holds energy, as if on deposit in a bank. That energy, which is called **latent heat**, is released when the water condenses back into fluid droplets. This heat changes the fluid dynamical movements in many complex ways. This is the crucial power mechanism in thunderstorms and, notably, hurricanes, but it also happens in smaller, more mundane settings too.

The vapour changes to liquid water all of a sudden, when the thermodynamic conditions require it, releasing the energy into the air movements. One of those thermodynamic conditions is how moist the air is, i.e., the relative humidity. So it can happen that the heat of some filaments is released while all around them, in the dry air, no heat is released.

To do the fluid dynamics correctly now requires, in addition to the basic (impossible) turbulence problem, tracking these filaments in detail within the flow, including the thermodynamics of whether or not latent heat is released along their length.

We can't begin to do this, let alone calculate in principle where exactly the filaments should be at any given time. These filaments will form up like an invisible ball of Devil's yarn, tied in demonic knots of filaments within filaments, rapidly changing, merging and separating. And here, you may recall, we introduced a simplification that there would be only one type of filament. As the molecular diffusion eventually spreads out the filaments, some of the dry air begins to moisten. It too is then pulled like toffee into filaments, only these new filaments are between being dry and moist. They act like a different colour of yarn for the Devil's invisible ball, and will be spun into it in a redoubled level of fiendish complexity.

In the free atmosphere, we don't measure such things. We can't measure such things. That is because we can't normally measure moistness detail on scales much smaller than a hygrometer (a moistness measuring device). But the measuring area of a hygrometer is larger than the thin filaments, so all it can do is give a type of instrument-averaged relative humidity over time.

Nor are such data directly useful. The thermodynamics in the filaments do not operate on averages. The heat is released when the conditions in the filament are right, not when some instrument-averaged conditions are right. The two theories call for all the detail when connected. There is a place in them to put every piece of it, but most of it is unavailable. They need it all. No global measuring network will ever achieve this. If you miss the tiniest piece of it, you have to make up something to stand in its place. You can't do calculations at all unless you make things up. It simply won't work without everything. Will the small effects from what we can't measure or make up make a difference? Can we get away with leaving out some of the information the theories call for? If you think it won't, how can you know?

The oceans have a turbulent layer too, with something like a Devil's ball of yarn in them. The sizes and time scales are different, but the idea is essentially the same. There the filaments are made up of saltier water, which leads to differences in buoyancy, leading to complex changes in dynamics. Small differences in buoyancy can lead to big movements.

But there is more. When water moves violently, pressures can locally drop and the liquid can turn to a gas. That change is a thermodynamical one set up by the fluid dynamics. This phenomenon, which leads to bubble formation, is called cavitation. It seriously complicates the basic fluid dynamics, even if we knew where, when and how big the bubbles would be. The fluid dynamical equations valid in the liquid no longer hold in the bubble. However, in cavitation, clouds of bubbles can emerge: big bubbles, smaller ones and smaller still.

The complex motions in the atmosphere and ocean are linked: The atmosphere gets much of its moisture from the ocean, while the ocean is driven by (turbulent) winds that rub against its surface. The interactions between the air and oceans form a whole universe of impossible complexities of its own. The resulting surface waves throw water more easily into the air than thermodynamics alone might suggest, and they swallow air into the water, which is dissolved or persists as bubbles. Where, when or how many of each type there are can never be precisely known.

We have no theory for that, and no way to observe it. The fluid dynamics and thermodynamics together place such impossible demands on us that we can neither measure nor calculate from either of these two classical theories alone or together. It is a kind of synergy, wherein the enormity of the impossibility is more than the sum of its parts.

Maybe we can hope to overlook some of it. But what parts can we afford to ignore? How small is small enough? No one really knows.

Let's now add in another layer of theory. We don't get to deal purely with fluids. Solids get suspended in the water and air. It would be so much simpler if they would just all fall out and we could be done with them, the way everyone expects. But it doesn't work that way. There are megatons of small, extremely slowly falling objects in the air, too small to see individually, that take weeks, months or even years to fall to the ground because they are so small. Of course, at the bottom limit of size, the objects are just molecules and not really small chunks of material. Between that very small molecular limit and what you can actually make out with your naked eye is yet another whole world of natural but insurmountable complications for us over a wide range of scale. The largest piece is 10^4 times the size of the smallest. We call these ubiquitous pieces of suspended material **aerosols**.

Aerosols are made up of nearly every solid material (and liquids too) that exists, including living things. They come in all shapes and they can clump together like invisible floating dust bunnies. The art of measuring their presence is not a trivial job, as they seem to be virtually nonexistent, but everyone understands that if a closed room is left undisturbed, over time, surfaces can become quite dusty as aerosols begin to settle out of the air.

Aerosols get embroiled in turbulent air movements. Moving air can draw them along and even embed them into the flow within filaments of the Devil's yarn ball. If they are busy falling when they get pulled back up, it's like a backwards game of Snakes and Ladders where penalties hurl you upward instead downward. They can be riding in a filament, and then, as the filament turns, the denser aerosol doesn't make the turn, crashing into the next filaments instead, crossing what's in between.

Aerosols, solid as they often are, typically carry a cargo of liquid water, which forms itself into a coat, bound on the outside by a smooth balloon-like surface created by its surface tension. Surface tension is what makes water drops behave like water-filled balloons. It arises because water molecules organize themselves only on the surface, due to the remarkable asymmetric way the molecule itself is put together. The water attaches itself to the solid in an amount determined by the relative humidity.

Think about the fluid dynamical and thermodynamical complications that many such droplets introduce when drifting across humidity filaments in a turbulent convective flow, evaporating and absorbing water as they drift across different filaments. Here we have solids, liquids and gases all coming into play simultaneously. Think about how the fluid motions affect what happens to the aerosols. This would be an impossible problem to set up even if we knew where every aerosol was, as well as all of their shapes and properties. We have no chance of doing that. Yet can we afford to ignore this detail?

As the relative humidity increases, the amount of water cargo can grow, until the aerosol is more droplet than solid particle. Collisions make these droplets coalesce into still larger droplets. If they become big enough, they can scatter visible light, forming hazes and ultimately clouds as the air reaches maximum moistness. If they get bigger still, their terminal velocity will grow and they will start to fall much more rapidly. As they fall, the evaporation rate may grow and they may lose their water, slow down, and then be blown back up again. As they evaporate water, they naturally draw the heat out of the surrounding flow that they previously put into it— latent heat works both ways. They could also become very large, accumulating water after complicated, turbulence-driven movements and collisions, until they become monster aerosols that fall out as the normal-sized objects we know as rain.

Aerosols are an especially nice example because much of the Big Panel's argument that carbon dioxide is a threat to the environment hinges on a particular view about how aerosols act on microscopic scales in the atmosphere. We will take this up further in Chapters 4 and 6. For now, we note that no conclusions about the role of aerosols in climate are founded on the basic theories from classical physics.

Rain changes the fluid dynamical conditions so strongly that it can shut off the upward movements that make a thunderstorm possible. So for the in-principle calculations, we need to know about all the details of rain formation. Imagine theoretically following every drop of rain everywhere in the world, and the airflow around each one, as well as the dynamics and thermodynamics on each drop's surface and within its interior as it falls. Add the case where the rain is frozen too. We would follow the growth of hail and the formation of snowflakes. In doing so, we would answer as a byproduct the age-old question of whether two snowflakes are ever exactly the same. That is what the classical theories require of us in principle if we do not want an incomplete theoretical treatment. We would need to compute every twinkle of every star (twinkling is caused by turbulent air movements) to carry it off in principle. What part of all of this is so small that we could leave it out?

The positions and physical details of the cloud droplets affect the dynamics, not just because of the direct consequences of the droplets embedded in the flow, but also because their positions collectively determine whether the flow is shaded or not. The energy from sunlight plays a role, but so does the invisible glow coming from the air below and all around in a form of light we call infrared. A cloud seriously changes the balance of energies going in and out of the flowing air, leading to net heating or cooling, which can significantly change the fluid motions, not to mention the temperatures and relative humidities, as well as the movements of the aerosols, droplets and rain.

The classical theory of **radiative transfer** treats the movements of energy in the form of electromagnetic radiation that we usually encounter as light. It is not the kind we know from 1950s science-fiction movies that feature attacks by some giant ants created by radition. It's just light—and the radiation we see is only a part of it. There is also light that is too red to see (infrared) and light that is too blue to see (ultraviolet).

Radiative transfer is another classical theory from the early 20th century. It is also a complete theory in the sense of Navier-Stokes, but it is more straightforward in many respects because it is linear and finite-dimensional. However, the complexity of it comes in not only because of the complex way molecules absorb and emit radiation, which we will discuss later, but also because of the way dust and droplet aerosols scatter light. It would be nice if the aerosols would co-operate and come in the same shape and size, but they are different sizes and shapes and are composed of different materials. Water attached to them interacts with their shape and the materials involved, adding to the complexity of how they scatter light. There is no general theory within radiative transfer for how light is scattered by all types of aerosols, although there are a few famous special cases. There are databases describing many known types, but basically the actual rule for scattering is often distilled from observations rather than computed from theory.

If this were not already complicated enough, the position of the scattering aerosols, not to mention the molecules in some cases, is determined by the thermodynamics and fluid dynamics. So the Devil's ball of yarn begins to affect the radiative transfer problem too. They feed back on each other. Which causes which? Where exactly are which types of droplets going to be to do their scattering?

We see droplets on a large scale as clouds. From a distance, they have structure, and as you examine many in more detail, you see smaller structures and so on. It never really smooths out, no matter how closely you look at them, assuming you don't go all the way down to the size of droplets themselves. When something doesn't smooth out, no matter on how small a scale you look at it, it is often best described by what is known as a **fractal**. Fractals are mathematical objects that are structured so they have a dimension that is typically a fraction. When you first hear about such a thing, you might think it's a joke, like having two and a half children, but when you look at the details of the idea, everything checks out and it makes sense. People have tried to represent clouds as fractals to do the radiative transfer calculations as accurately as possible.

Fractals are well established in the history of turbulence theory and chaos too. The solutions of the Lorenz equations traced out a fractal, and the Devil's ball of yarn is a good candidate for this concept as well. It has been proposed that the singularities in the vortex picture of the solutions of the Navier-Stokes equations be arranged as fractals. These old theories are being freshened up with new ideas to cope with the new insights.

We could talk about how the oceans interact with the land, the exotic thermodynamic property of ice compared to other solids that makes it possible to skate and ski, and also for glaciers to flow, making them the subject of wonderful mystery and unpredictability. We could talk about the land-surface-air interactions, and how we not only have to think about the flow over land of different heights, but around buildings, through forests, past every leaf. We could talk about the ever so important first kilometre of air above the ground and ocean, and all the rich chemistry that goes on in the air and ground; the gases emitted by the soil and volcanoes; the gases absorbed and lost; the chemicals that the rain cleans out of the air; the fluid-solid interactions of rivers. And we have not even arrived at butterflies or seagulls, or the family dog, for that matter.

Butterflies or seagulls are such wonderful and whimsical beginnings to the failure of predictability. It is absurd to imagine that these marvellous creatures could ever be reduced to theoretical predictability, let alone that we could predict their every action or follow each nuance of the movements of their wings, forcing the fluid with every stroke. A theory for butterflies is external to our network of classical theories, even though it doesn't undermine Navier-Stokes theory itself.

Navier-Stokes theory itself actually fails at times for fluids in the atmosphere, with consequences far greater than the stroke of butterfly wings. When a thunderstorm erupts, the electric charge differences between places within the air and between the ground and the air are reduced through lightning. In the stream of the lightning discharge, there is another state of matter called *plasma*, where the Navier-Stokes equations do not apply. The discharge creates a shock wave, which makes thunder, that also undermines Navier-Stokes theory. Lightning is governed by classical electromagnetic theory. Like Navier-Stokes, it is regarded as a solved problem, but no one knows from it precisely when a discharge will take place, nor precisely what the path of the discharge will be, let alone every detail of the plasma that is created.

Yet each of these events changes the circumstances. Think of the thousands of thunderstorms going on in the world at this instant, hammering out a rousing global chorus in the initial-condition-error polka. Never mind the graceful and continuously moving butterfly, or all the rest of the other distinct impossibilities we have faced here, for that matter. To use Navier-Stokes theory exactly, you have to reset and restart your calculation after every single stroke of lightning. So it isn't just an initial error; it is an ongoing string of abrupt, discontinuous restarts.

If you want to do the in-principle calculation, you have to know what we do not know theoretically, such as when a lightning strike will take place. Then you have to measure things that we cannot measure, such as the relative humidity everywhere, and put it all together to compute something that is too big for the biggest computers we have ever realistically imagined. And after the next lightning stroke, you have to start all over again from scratch, unless you plan on an in-principle plasma physics plus shock wave calculation too.

We do not know how small the detail is that we can leave off to adequately understand climate. We hope that some of it will cancel itself out in some grand averaging scheme. So even if we cannot forecast the weather, maybe we can forecast the average weather. That is the hope, since we are clearly not up to the task of directly facing the sublime complexities we have only touched on above.

However, once we start saying we will throw away detail because we can't handle it, because we don't have it or because we don't know what detail we really need, we are not doing theory anymore. When we start to make up the missing data and the missing theory, and define what is too small, not by what we have shown to be physically too small to matter, but by what is as small as we can manage, it is not theory anymore. It is modelling.

Models in Lieu of Theory

Models may run on computers, but they are not the product of computer technology. Computers, powerful and mysterious, give a certain cachet to modelling, but all that computers really add is the ability to carry out the requirements of a model with tireless and mechanical precision at a very high speed. As far as a computer is concerned, a model is just like a theatrical script the computer acts out with inhuman precision. The model is the script. If the script is bad, the computer adds nothing to it, at a very high speed.

Some models are not mounted on a computer at all. The Lorenz equations we discussed in the preceding section constitute a model of turbulent atmospheric flow. The equations themselves are the model, although a computer is required for us to see the solutions. Incidentally, it is only in the last few years that we have been able to prove that the chaotic solutions of the Lorenz equations actually are what the computer has been telling us they are. For nearly 40 years, we were not completely certain that they actually behaved the way the computers had been saying.

Why did we doubt the results? For a very good reason: Computer calculations can produce chaotic behaviour that is not present in the equations the computer is being used to solve. There are some infamous examples, such as a computer treatment of the logistic equation. Some years ago, Chris and his colleagues introduced a case where the computer calculation kills off chaos that is actually present in the equations.

Computers make mistakes. And they make them very rapidly. You can make errors bigger and faster on a computer than in nearly any other way. You can get a sense of the errors that are produced because of the finite size of computers by filling in the following table for yourself with what your favourite computer or calculator gives. (The Microsoft calculator program doesn't use standard numerical techniques, so this won't work for it without an adjustment, which we will not go into here.) We have filled the table in using an important mathematics program for computers called Matlab to help the flow of the discussion, but we recommend that you actually try it yourself. The only thing that may differ slightly from machine to machine is where the answer switches from the right one, 1, to the wrong one, 0, as you fill in the table for yourself from top to bottom.

Too many people have an unrealistic faith in computers doing mathematical calculations. The fallibility of computers in mathematical calculations remains widely unknown, and there is tremendous resistance when it is pointed out.

The most fundamental limitation they have in terms of theories and their accompanying mathematics is the finiteness of their machine precision. This manifests itself in many ways. One of these pertains to the scripts we can write for models. We can't write down just anything in mathematics for the computer to do. Any instructions must take into account the finiteness. Mathematically it means that some expressions cannot be used and that we are often forced to employ approximations wherein the smaller bits of the mathematical expressions are pruned off.

Table 3.1. The Fallibility of Computers	
Input	**Output**
(key in exact sequence)	
$100,000^2 + 1 - 100,000^2$	1
$1,000,000^2 + 1 - 1,000,000^2$	1
$10,000,000^2 + 1 - 10,000,000^2$	1
$100,000,000^2 + 1 - 100,000,000^2$	0
Etc.	0

This is called truncation, and it produces truncation errors. In many cases, if you know what you're doing, it doesn't present great difficulties. But it can be a problem when small things can have big effects. That is why it took almost 40 years to be convinced that the calculations on the Lorenz equations were okay. And it is one reason why we cannot do turbulence calculations generally.

Pruning equations can be done in many ways, and differing circumstances demand different approaches. Sometimes you know how to write approximate equations down as accurately as you want at will. All of the respective levels of truncation are laid out systematically in an infinite series, which you can just chop off at a level of accuracy where you are satisfied. If things don't work out, just add more detail by turning to the next terms in the series. After adding in that detail, you can test it to see if the detail is sufficient for your task. If it isn't, add more. This will always work with a big enough computer, where nothing like chaos or turbulence is happening.

Of course, in the presence of turbulence, things are not so simple. Nor are they if you do not know how to, or do not want to, write down all of the equations. Then it is not so clear how to add in more detail to get a satisfactory result. Just any old detail in any old way will not do. Only the right structure for the detail will lead to a model with predictive powers. You can only get that structure from the theory, or, barring that, by doing controlled experiments.

Maybe the model is not something that needs to have predictive powers. A case in point would be the dynamics of fluids in movies, where artists and computers have created artistically satisfying dynamics. Before computers, the classical Disney movie *Fantasia* had many examples of wonderful dynamics painstakingly drawn by the hands of artists, frame by frame. Those dynamics included the movements of water. Cartoon dynamics are as much an art as that of the individual still images. But, at the same time, the movements within the movie constitute a model. This model, though, cannot tell you what the water will do afterward, any more than it can say what Mickey does after the movie. It has no predictive powers.

It is important to mention a pre-computer fluid caricature like that in *Fantasia* because computers have the capability to add in so much more detail than what was formerly possible only by hand—so much so that model movie fluids are now easily confused with theoretical calculations. The cartoons don't need to look like cartoons anymore, even if that is what they really are. They are even being used as special effects instead of just as cartoons.

In the movie *Shrek,* there are some delightful fireballs from a dragon that do not immediately seem like cartoons because of such computer modelling techniques. But wonderful as they are, you cannot predict with them, any more than you can with the cartoon water in *Fantasia.* They begin with basic structures like balls and cubes and then add in layers of artistic detail and effects. The choice of detail gets added in an artistic matter, based on what the artist likes best. They do not solve the Navier-Stokes equations in *Shrek.* It would be foolish for the artists to take that approach.

For *Star Wars Episode I,* cartoon water was used as a standin for real water in one of the missing scenes that has made it onto the DVD. There, a waterfall is revealed to be not water at all but falling sand that is altered by similar sorts of computer artistry. No Navier-Stokes here either, but it looks real enough. It is pure, wonderful, magical artistry—and you can't predict anything with it.

All models have some form of artistry in them. They are something between science and art. Some are more science while others are more art, and there is everything in between. They are subject to the rules we can manage, and the rest is filled in with experience and an aesthetic flare for how the rest of the thing might go. The result is guided by a pragmatic sense of whether the made-up parts actually fit the way they ought to. For the animator, the pragmatic test is whether it will delight or fool an audience, and for the forecaster, it is whether those adjustable bits of the model can be adjusted enough to agree with experiment.

A great deal of industrial mathematics involves models that employ what we call empirical relationships, also known as parameterizations. Such relationships are established by experiment, with little or no theoretical basis. It is clear that these empirical relationships, in a suitable controlled experiment, will be a valuable aid as part of a model, if the model is developed to work only in conditions over which the experiments have been done. In such a context, predicting or accurate simulation can be done very well. Moreover, for more detailed forecasts and simulations, more detail can be added and confirmed experimentally. Models and experiments work hand in hand. But if you push an industrial engineering model outside of the domain where experiments are done to validate it, you may void your insurance.

Meteorological models are filled with parameterizations too, but they have many different issues. The empirical relationships can be tried out and refined, again and again. One of the distinctive problems in that field of modelling, other than the search for the best empirical model relationships, is that the models call for more starting data than we can actually provide. You can't run them without the data, so you have to make the missing data up.

But not just any old data. It has to be something that does not introduce obviously absurd behaviours. Some models take more time making the missing data up than calculating the forecasts. The rules used to make the data up are themselves a form of modelling and also represent a kind of artistry. Meteorologists look at what the models finally do and the original data directly to make the official forecasts. Unlike industrial models, meteorological models are not really good enough to do it all by themselves. The meteorologists are like the animators in that they adjust the product to reflect their repeated experience of what works in their respective fields.

Meteorological models are not climate models, although some mistake the latter for the same thing run a bit longer. Climate models focus on different processes, and they must capture the full range of time and space scales that the meteorological problem can effectively ignore. We repeatedly emphasize, no one can do, nor ever has done, controlled global climate experiments, and systematic observation does not constitute controlled experiment. Unlike meteorological models, climate model parameterizations have not been tuned after repeated experience with climate change. Moreover, unlike meteorologists, no climatologist has lived through repeated events in his or her field so as to acquire a personal sense of experience of what to forecast. How many ice ages have you lived through?

Meteorologists can make reasonable public forecasts of the weather, but climatologists can't really forecast climate similarly. This is not a criticism of the field; it is the reality of it. No one faults cosmologists because they cannot tell definitively whether the universe will expand forever or collapse back onto itself (collapse is in just now). The field is scientifically rich and humanly interesting without this particular ability. This is true for climate research too.

Of Mice and Models

Climate models come in some broad types that are characterized in what has come to be known as the hierarchy of models. The idea was that as one introduces more detail, models improve, becoming superior to their less detailed cousins. Averaging comes in here too, in all of its wild, free-form climatological ways.

In this sense, the simplest climate models are "zero" dimensional (i.e., 0-D). This usage refers to the number of space dimensions included in the model. Obviously there can't really be zero dimensions or there wouldn't be much to talk about. Such models do have degrees of freedom in that they can have a single temperature, or maybe some winds (amazing, with nowhere to go, but it checks out; it's a model!). Such models represent how the Earth behaves collectively, in cold space, while irradiated by the sun under a variety of circumstances. Zero-space dimensions are traditionally regarded as the product of some kind of averaging process. Nobody ever says how to get from the full atmosphere to a 0-D model by averaging. It is really just heuristic, and the model is just a cartoon—a useful cartoon, but a cartoon nonetheless.

Moving up the hierarchy, you find the 1-D models. These are also mystery-averaged models. Of course averaging suggests that there is less detail than there would be without the averaging. But this is misleading because it implies that you lost detail you never actually had. We don't know how the two space dimensions are eliminated, and we wouldn't really want them if they were available in the first place. Such a model is meant to keep things simple. In one class of 1-D models, the average is some sort of average across longitude and the vertical, resulting in just one space coordinate between the equator and the poles. Things often get flattened too. So the whole planet is like a big long slab of iron heated at its centre (the equator) and cooled at its ends (the poles). The most famous members of this class of models really worked that way too: like a slab of iron. There is no fluid dynamics in it at all, and the radiative transport is minimal.

The other class of 1-D models has a historical connection to astrophysical models. This class represents only the vertical dimension. The horizontal is eliminated in the usual heuristic way. The physics in vertical in the first instance is in terms of rough numerical computation of the equation governing the theory of radiative transfer, called the equation of transfer. The radiative transfer theory is accommodated by assuming that the world is flat and that the atmosphere is made up of uniform layers. There usually aren't very many of them considered. Then the temperature in these layers is computed.

The first sticky modelling points here have to do with what you provide to the radiative transfer theory. Do you consider clouds, for example? Clouds are created physically by everything that is "averaged" away. So if you want to consider them, you have to "fudge them in" in an artistic manner, unless you want to do the whole problem first before you average—not much point in that. You can't have your cartoon clouds varying with direction; otherwise, you will not be doing a 1-D model. They have to be uniform. It's a model.

With your cartoon clouds, do we believe they will vary with the model conditions in some manner? What manner? People have got into big disagreements about this, but there is no theoretical reason for choosing one over the other because we are dealing with a cartoon at this stage. It is like two animators arguing whether the cartoon mouse should have a bow tie.

Do other things needed by the radiative transfer theory vary? The modellers decided yes, so the tradition developed to vary the amount of water in the atmosphere, based on the temperatures. A hand-waving argument can be made for this, but thermodynamically, water amount depends on more than that in the real climate. And the other stuff it depends on isn't in the model. So what you decide it should depend on doesn't come from the physical theory. It's like the waterfall made from sand—food for thought, but not a basis of action.

There is more, but let us jump to the last and most problematic part of them all. The temperature of the atmosphere is not determined by radiation alone. That is why we discussed fluid dynamics so much above. The lower 10 to 15 kilometres tends to be convective, although not everywhere and not all of the time. Convection means turbulence. The convection air conditions the surface. It removes energy from the surface as well as the radiation does, so you cannot ignore it, even in the model. If it were not present, our world would be like the inside of a real greenhouse—the temperature where you live might go as high as 60°C.

You can't calculate turbulent fluid motions in a 1-D model. The fluid would go up, but it could not go down again without passing through itself. Furthermore, in the real atmosphere, some places have different kinds of fluid movements with different properties based on the circumstances. The model will allow only the one case. That is only fair. After all, the model allows only one surface temperature, which we know cannot be correct: There are many surface temperatures on the Earth. We can say that the temperature is an average, but average over what and in what way? The surface temperature is a cartoon in one degree of freedom. There is nothing wrong with that. It is a model.

There was once a big debate among some modellers as to whether the surface temperature of the air should be different than the surface temperature of the ground. If it were, then there would be two "global" temperatures: one for the ground and one for the air at the surface too. The pure radiative transfer calculations say there is a difference, but it is conventional to say that the air movements eliminate the difference. Are they sure that this can be meaningfully discussed in the context of the model? Would it make much of a difference to the behaviour if a model difference were inserted? Don't count on the physics to tell you what to do with a cartoon. There is no physical basis to clarify the issue.

Of course the air movements significantly change the movement of energy in the atmosphere. Once the modelling rules are put into place, the resulting radiative-convective model is no longer computing anything like known equations of basic theory. It is a model. These rules are so traditional (handed down for generations) that many, even on the inside, do not know they are just made-up rules that seemed nice, and nothing derivable from the basic theory. But the global temperature is always a nice and comfortable global 15°C in radiative-convective land, if you don't do a global warming experiment.

We can put in more detail and join up the different 1-D models to get 2-D models. You can imagine what might be said about that. But we need to move right on to 3-D models, because these are at the top of the modelling food chain. They are the ultimate in terms of modelling climate detail. They are the great hope of those who do not know models or modelling, because they think that allowing three dimensions amounts to throwing in all of the science. People mistake these 3-D models, known as general circulation models (GCMs), for the ECMs of the previous section.

But they are not ECMs—not even close. They have much more detail than the models lower down on the hierarchy, but, as we warned earlier, more detail does not imply that your modelling is converging to accurate behaviour. It is a common fallacy that by imagining a limit, you can automatically take any process and you will go to the limit you want. Mathematics has many hard lessons for people who think that.

To see this point, instead of climate models, let's consider a hierarchy of models of a mouse, as illustrated in Figure 3.1. With each new free dimension added, there is more detail and the model improves. But in the 3-D limit you do not have a real mouse, you have only the 3-D model of a mouse. Everyone can see that the result is a representation of a mouse, but everyone can also see that it is not actually one. That is what models are, caricatures.

The atmosphere and oceans are not three-dimensional, they are infinite-dimensional. Space dimensions are not the single crucial issue in achieving accuracy. Merely creating a 3-D model and pumping in all the detail you can manage is not an ECM.

This is not to say that GCMs are Mickey Mouse. Far from it! They are truly magnificent creations among models. The best of them have been in constant development since the late 1960s and generations of modellers have been involved in their construction. We are speaking of hundreds of person-years of labour in these models, because they are developed in teams. The details in the creation of one of these is more than can fit into any one head, as it takes specialist knowledge in a variety of subjects to do justice to the various components that go into them.

GCMs have given much insight into many things. Aside from their

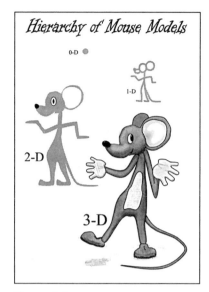

Figure 3.1

immense academic value to the modelling and climate community, they were the first to give a convincing connection between ocean temperatures and seasonal weather. This has led to a revised approach to weather forecasting, although it is not really climate forecasting. They also constitute our best computational estimates of climate change. Because of them, the notion of climate change is plausible, and not something that can be casually dismissed. Finally, they have awakened our awareness that the issue is a complex one not done justice by the idea of a "global temperature."

As human creations, GCMs are to be admired, but people who do not understand what is not in them tend to get carried away. They are treated by many as ECMs. Studying GCM model outputs for the purpose of forecasting has become an industry. So not only do we have climatologists studying the climate, we have people unversed in the modelling content studying the climates of GCMs for the purposes of making forecasts.

Even though such modellers can be viewed as the no-nonsense gurus of the climate field, their caution about the nature of models is often not heeded. It is surprising that you can pick up anything at all on this in the PUN reports, where debate and disagreement are so consistently absent, but it is all there if you read between the lines. For example, a debate was cited wherein some of the output modellers wanted to use data from individual grid boxes to construct their climate scenarios and the modellers apparently objected. So what was that all about?

Grid boxes are to GCMs what pixels are to computer or TV screens. A pixel is the smallest dot that makes up a picture you watch on the screen. The PUN report indicated that modellers do not consider their models useful on the scale of a "grid box." Extracting climate information from an individual grid box is like watching a TV program on mute from one pixel. You might have some trouble figuring out what the program is about.

One complaint about GCMs by those anxious to make forecasts is that there is not enough detail in them. These grid boxes are hundreds of kilometres or more across. This is another manifestation of the finiteness of computers. There has to be a cutoff of the detail that will be considered by the model. There is just not enough space in the computer to consider more. It is finite. By the time you put in all of the details, oceans, air, parameterizations and other physically based structure, there is just no room for more.

GCM enthusiasts do not complain that there is a lot of missing physics in models, they complain that there not enough detail to talk about specific climate forecasts within scales less than hundreds of kilometres. So regional circulation models (RCMs) were created as models that filled in details fed to them by the broader inputs from the GCMs. This is the Kolmogorov trashcan in models: Big models dump their data to smaller ones, but not the other way.

While this is an interesting academic modelling exercise, it is another case of the fallacy of increasing model detail. The RCM cannot forecast climate with the missing detail that the full GCM does not contain. What it will do is embroider the information in the GCM, as does the Regional Mouse Model in Figure 3.2.

The additional detail this model will produce will not be available for the GCM; the flow of information is one way. Small things will not have any big effects in this pair of models. There will be only an RCM in a region of interest. If there were RCMs everywhere, you would have defeated the limit imposed by computer resources, so that doesn't happen.

Modellers know about the problem of small things having big effects. They call it the problem of sub-grid scale phenomena. The real trouble is that nearly everything important about global warming (radiation, turbulence and aerosols, to name a few) are sub-grid scale. In fact nearly the entire problem in the vertical is sub-grid scale. This is fair enough, as the practical vertical dimensions of the model atmosphere must be orders of magnitude smaller than the horizontal resolution of the models. This presents a difficult problem for modellers.

It means that what is done vertically is heavily done with the made-up rules, which we called parameterizations. Everything from convection to clouds, rain and the general cycle of water into and out of the system— everything that has to do with moving energy from the surface of the

Regional Mouse Model

Figure 3.2

Earth to space—is made up. This is not bad. GCMs are just models. It's in the acronym. That's what the M stands for.

The radiation has to be dealt with in some kind of compromise between a direct calculation and one that gets stitched like a mismatched limb of Dr. Frankenstein's monster into the much larger scales of the horizontal problem. One scientist once said that GCMs are really two-and-a-half-dimensional.

Navier-Stokes theory doesn't really get used in them at all. In the vertical, no attempt is made to compute those equations, while in the horizontal, the resolution is so coarse that the issue of viscosity is meaningless. The only way the model communicates with the small scales, where viscosity does play a role, is through the parameterizations.

Despite leaving out and making up so much, GCMs are mathematically deeply complex, especially now that they include atmosphere-ocean models too (AOGCMs). When there are subtle problems, it can be a serious job to figure out how to fix them.

GCMs are notoriously unstable. There was an infamous problem for many years called climatological drift, wherein the model would not settle down but would gradually change to different behaviours or "climates." There were problems with conserving physical quantities, which is not so strange in big numerical calculations, especially ones with lots of parameterizations. When you try to get rid of these odd behaviours, how do you know that your fix is not getting rid of something that the model is correctly capturing too? Maybe real climate does drift slowly, even in the perfect calculation on an ECM. Of course such weird behaviours can result from the parameterizations, which do not really conform to the laws of physics.

Models of a single thing are also not unique. You can construct all kinds of caricatures of the same thing. Everyone knows what a good caricature is meant to represent, even if it may have many deviations from

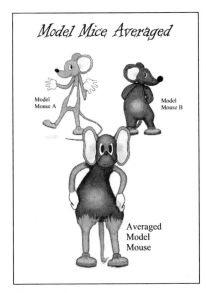

Model Mice Averaged

Model Mouse A

Model Mouse B

Averaged Model Mouse

Figure 3.3

what it represents. Everyone can tell that a cartoon mouse represents a mouse, and we can all tell that it is not a real mouse. But this difference gives considerable range in the possibilities of the model mice.

GCMs, like any other models, are not all alike. The GCM from one group behaves differently from that of another. If they were perfect calculations with an ECM, they would all be the same and they would behave identically. It is no surprise that there are deviations, as they use different internal (non-physical) schemes. The caricatures of the climate that they represent are all slightly different. But the GCM groupies who harbour the ECM dream still imagine that more detail gives a truer result. So instead of viewing the deviations of different models as an indication that ideal calculations are not being done, they see it as an opportunity to throw in more detail uncritically.

Just as there was the fallacy of uncritical detail in the case of the model hierarchy, there is one in terms of averaging across models. Maybe, the idea goes, if you take the outputs of various models, the additional detail summed up and then distilled by an average will be even better. Maybe, if we are lucky, the errors in all the models will tend to get cancelled out in an average. But what average? As we can see from Figure 3.3, in terms of mouse models, an averaged model mouse need not be any closer to the real thing than the models from which it was extracted.

Have you ever heard of the famous forecast of global warming that the "global temperature" would increase by $3 \pm 1.5°C$? Where do you think they got the uncertainty measure? It came from looking at the various models. There is no physically meaningful way to get such a number, for many reasons.

The future need not be in any of the models. We have no reason to say that any of them are correct. Unlike engineering applications, we have not validated the model parameterizations experimentally over the range of climates we want to investigate, although many people use terms like "validation" and "testing" of climate models to talk about observing what the Earth does now in our current climate. We don't know how to validate them properly. Averaging them need not get you any closer to the true future if it is another, different climate. So the uncertainty number on the change of "global temperature" is more a decoration than something with real scientific significance.

The same problem exists in studies where the opinions of climate scientists on the future of the climate were solicited and statistical methods were used to make a forecast from the range of opinions. Unbelievable as it may seem, people really have done that to forecast climate. It's even in the PUN reports! This curious approach is a model too, but the Earth doesn't care what any of us think. No climatologist alive knows for sure what will happen in the future, and there is absolutely no reason to presume that they have any more insight "on average."

We haven't said anything here that genuine GCM modellers don't know. We are not concerned about persuading them of anything. As the opinion studies imply, there is a range of opinions about the future, but they are agreed on the limitations of climate models themselves. They know about these things.

Measured against reality, GCMs are primitive and childlike, but they are the best we have and they are precious things. The problem comes in through those who oversell the models, inspiring inordinate confidence in what they are actually able to do. It manifests itself in headlines that report, with genuine astonishment, that part of the world is not acting the way the climate models "predicted." (If the papers really tried to cover that theme, they would run out of ink.)

The human problem is further compounded in the habit of reducing the output of GCMs to one degree of freedom in the form of "global" temperature. Think of all that goes into a GCM in terms of pressures, wind, and relative humidities, not to mention the enormous numbers of distinct temperatures. Think of the feat of computational science to pack all that onto the most powerful machines anywhere, and then hope that something sensible and stable comes out of the staggering complexity of it all.

But all that really gets mentioned is a single temperature, as if none of the rest of it matters. Ironically, GCMs are reduced to what many people really wanted all along: 0-D models. For them climate is just "global temperature," models are just fancy electric thermometers, and all the other physical variables in the atmosphere and oceans are just some other type of temperature too. And the cruelest irony of all is that "global temperature," as you know it, does not exist. Oh, you didn't know? Well then, pay attention, because the great Professor Calorie J. Thermos just walked in.

Professor Thermos Teaches a Lesson About Global Temperature

PHYSICS 250: ELEMENTARY THERMODYNAMICS. Cournot Hall, Room 311A. Prof. Calorie J. Thermos.

Thermos: Good morning, everyone. I apologize for being late. I got caught in the midst of some sort of student protest out in front of the administration building. Does anyone know what it's all about?

Student: They're protesting global warming, sir.

Thermos: I see. What's your name again?

Student: Nicholson. Crank Nicholson.

Thermos: Crank Nicholson, eh? So they're protesting global warming. Do they actually believe our administration controls the global climate?

Nicholson: Not exactly, sir, they want the university to sell its shares in oil and coal companies.

Thermos: Ah, I see. No, wait, I don't see. You'll have to excuse me, I don't read the newspapers very often and I'm not up on all these new theories. How will selling some shares change the physics of the atmosphere?

Another Student: Well, sir, the use of fossil fuels is adding greenhouse gases to the atmosphere, and this is causing global temperatures to go up. There's a scientific consensus that fossil fuel use is pushing global temperatures up, which will damage the environment. So they want the university to send a message to the fossil fuel companies that this has to stop.

Thermos: Oh, yes, greenhouse gases. Those things. What's your name?

Student: Runge Kutta, sir.

Thermos: Thank you, Ms. Kutta. Now, you said something about the "global temperature." Tell me, what exactly is that?

Kutta: Uh, it's the Earth's temperature, isn't it?

Nicholson: I have a graph of it here, sir. I got it from the NASA Web site. It shows the global temperature, and how it has gone up over the past century. You can see why everyone is worried about global warming.

Thermos: Let me see that. Well there it is. Son of a gun. That's quite something. But do you know what I find truly remarkable about this picture?

Kutta: That global temperatures are going up so quickly, and that the planet is warmer now than it has been for thousands of years?

Thermos: Nope. What is remarkable is the precision with which these folks are able to measure something that does not exist. Graph or no graph, there is no such thing as a global temperature. This is a global temperature *statistic*, but it is not a "temperature." The world is not in thermodynamic equilibrium, so there is no single temperature to discuss. What we measure is tied to places through local thermodynamic equilibrium. It has no global meaning, and some statistic certainly does not establish whether the world is warmer or cooler than it was 10, 100 or 1,000 years ago. There is no scientific way to show such a thing.

Nicholson: Local . . . huh? What do you mean, it's a "statistic," not a temperature? What's the difference?

Thermos: With anything that is measured numerically, you can, if you like, take a sample of observations, add them up and calculate some type of average. That's just doing statistics. But sometimes the thing being measured only means something locally and loses its meaning when added up or averaged. Here's an example. [Thermos writes seven digits on the chalk board.] I took the phone numbers of all my colleagues in the physics department, added them up and computed the average. Now, if you dial this number, does that give you the average physicist? Of course not. Telephone numbers only mean something individually, when attached to a single line. Add them together and they instantly lose their meaning. The "average" phone number is a meaningless absurdity. In the same way, "average temperature" is meaningless. Numerically, you can add up a bunch of temperatures and take some average, but it has no physical interpretation. Temperature only means something locally, because the thermodynamic conditions vary from point to point.

Kutta: But we could determine the average temperature for this room, sir. And if we did it today and again next spring, we could figure out whether it is warmer now or then, couldn't we? Why can't we do the same for the climate?

Thermos: Good question. Let's try. Mr. Nicholson, have a look at the thermostat and tell us what temperature it is set at. Meanwhile, I am going to put this laboratory thermometer on the window ledge here.

Nicholson: It is set at 20°C [that's 68°F], Professor Thermos.

Thermos: That's interesting. This thermometer here measures a temperature of 17°C. Which is the right temperature for the room, 20 or 17?

Kutta: Wait a minute, it's winter outside, we would expect it to be cooler by the window.

Thermos: So? Why doesn't the window ledge temperature count as much as the temperature at the thermostat?

Nicholson: That temperature is not representative of the room. It comes from near the window, not in the middle of the room. It shouldn't count.

Thermos: So you are looking for a "representative" temperature that would be inside some imaginary line between the middle of the room and the window? Would where I am standing now, at the desk in front of the class, be within this imaginary boundary?

Nicholson: Sure. [General agreement is expressed.]

Thermos: All right. [Thermos holds up another thermometer.] I'll just put this into my glass of ice water here like so. [Commotion, gasps and squeaking chairs as Thermos watches the thermometer reading relax to the temperature of the ice water.] So, according to your rule, it is 4°C in here. That is pretty chilly for a classroom. And relating to your question, Ms Kutta, if you watch this classroom temperature over the next hour, I am confident that it will change, upward, I believe, despite the presence of a thermostat. So the room will prove to be "warming."

Nicholson: The ice is just melting and the temperature of the room is not 4°C. You didn't say anything about ice water. That was a trick.

Thermos: I was just following your rules. Are you saying that I should measure temperature in a different place than this?

Nicholson: Yes.

[Thermos pulls out yet another thermometer and drops it into the glass of ice water too. There is puzzled silence.]

Thermos: Well, the new thermometer is in the glass of ice water too, but it is in a different place within it, which is a different place within the room too.

Nicholson: Let's just exclude the entire contents of the glass and any other tricks like that.

Thermos: Okay. [A fourth thermometer appears and is deftly thrust under Thermos's tongue. There are bleak looks from the class.] *Vfhat! Thill not Thafithfied!*

Kutta: That doesn't count either! It certainly isn't 37°C [98.6°F] in here. I think we need to measure the air, free of heat sources and cold places like ice water.

[Thermos takes the thermometer out of his mouth.]

Thermos: First you want to exclude a region near the window, then you want to cut a hole out of what is left by excluding the glass of ice water. Now you want to exclude all the space occupied by the people in the class. The classroom that you want to measure is starting to look like a piece of Swiss cheese. Where does it end? Do we cut out a region near the radiators too? How about near the classroom door? What about near the lights? How about near the wall that is being warmed by the sunlight? How big do you make these cutouts? Do we cut out regions around the currents of warm air from the various sources in the room too? What about the currents of cold air?

There is no place in this room that is not affected by heat or cold to a greater or lesser degree. Listen carefully. The matter at every point in this room—by the window, by the radiator and within each of us—has a temperature of its own. There is no one "temperature-of-the-room": There is what we call a *temperature field*. It consists of an infinity of temperatures, even for this small room. Most of the field's temperatures are close by each other, but there is electromagnetic radiation in the room too, and it has temperatures of its own that are quite different. The sunlight has temperatures that are many thousands of degrees. Some radiation, from radio stations and cellphones, passes right through us with little effect even though it has temperatures of its own too. The air is certainly not thousands of degrees.

Why can so many widely different temperatures exist in the same place? This room, the air, the sunlight and all of us are not in thermodynamic equilibrium with each other. If we were in equilibrium with the sunlight, we would be a disjoint bunch of small molecules and ions. The air is less drastically out of thermodynamic equilibrium with itself than with the sunlight, but it too is out of equilibrium with itself. The only time there is "one" temperature is in thermodynamic equilibrium.

Nicholson: I thought that temperature just meant energy. So what does 5,000 degrees mean if you can't measure it with the thermometer and it doesn't cook you?

Thermos: Thermometers work by going into equilibrium with what they measure. If you could trap enough of the radiation and stick the thermometer into it, it would register that temperature, but it would probably melt first.

As for your first remark, temperature is not energy. That is a common misconception. If it were, we would not use temperature at all, we would use energy instead. It is better to think of temperature as indicating the condition of a physical system in terms of how energy is distributed over physical states. Remember, energy can take many forms simultaneously in physical systems (kinetic, heat, chemical, potential, etc.), and out of equilibrium, it can be arranged in specialized ways. If you distribute even just a small amount of energy over even fewer states, you can have a very high temperature. A laser pointer is a good example. Its temperature is in the tens of millions of degrees. Some thermodynamicists will approximate the temperature of a laser as infinite for some practical problems. And all of that temperature comes from a couple of flashlight batteries! You can't even feel the warmth from a laser pointer on your hand! No. Temperature is not energy.

Kutta: When people talk about global warming, they say the global temperature is going up. Everything you said applies to the environment outside. Are you saying there is no such thing as global temperature, and if there isn't, does that mean there is no global warming?

Thermos: You asked two questions, Ms Kutta: Is there a single temperature for the Earth and is there really global warming? The answer to your first question is that *there is no such thing as a global temperature*, precisely as there is not a single temperature for this classroom. There is a temperature field, a continuum of temperatures, that cannot be reduced in any sensible way to a single number. The answer to your second question is a longer story. Some people go ahead and reduce the temperature field to a single number by inventing a kind of average that they like, and then they see if it is increasing or not.

In terms of the range of temperatures that exist in the world, the change they are looking for, this has little physical meaning. That doesn't mean that we aren't having global climate change, it just means that many people who are looking for it confuse their particular average of thermometer readings for climate. There is a great deal more going on in the environment than just changes of temperature, let alone some number that is a made-up rule applied to the temperature field. It is not a temperature of anything at all; it is a statistic.

Nicholson: I am confused. You talk about "an" average, and a "made-up rule," but the average is the average. Isn't there only one average? Why wouldn't it have physical meaning? Wouldn't an increase in the average mean that it's warmer? If the average height of the people in this classroom increased, that would mean that people are generally taller. We can say that without misunderstanding. Why can't we say the same thing about temperature?

Thermos: There are lots of different types of averages. The mean, where you add up the values and divide by the number of values, is the most popular one, but certainly not the only one. Two other favourites are the median and the mode, but there are infinitely more when you start to look into the real possibilities. All averaging really does is replace a collection of numbers with a single number. Ideally, the resulting number should represent the collection in some way, but there are infinitely many possibilities.

To be physically meaningful, an average has to be connected in some manner to the physical problem. In your example of average height, a changing average doesn't mean anything about an individual's change in height, of course. But if the average is, say, a mean, a growing average means that the length of the whole class, lying down head to foot in a line, grows. The change in average connects directly to the change in the overall length of the class as a whole. You can't do this reasoning with temperature, because "the total temperature" doesn't mean anything physical. It's just a bunch of numbers added up.

Nicholson: Why can't you add up the temperatures? For instance, we could divide the room up into thousands of one-centimetre cubes, take the temperature of each, including the people and the ice water and everything, then add these up and divide by the number of cubes. Isn't the total temperature a temperature too? What does the total have to do with the physical meaning anyway?

Thermos: The total is where you construct a direct physical connection here. For example, if we talk about mean energy increasing, this means that the total is increasing too. Increased average energy means something physical. Temperature, in contrast, is not an amount of something. It is a number that represents the condition of a physical system. In thermodynamics, it is known as an "intensive" quantity, in contrast to quantities like energy, which have an additive property, which we call "extensive" in thermodynamics.

An example of something that behaves intensively would be the percent of milk fat in a coffee creamer. If you put two small containers of 10% coffee creamer together, you do not get 20% milk fat. The cream is still 10%, even if you have twice as much. In the same manner, if you have two identical boxes with the same energy and the same temperature, join them together. The resulting doubled box will have twice the energy, but it will not have twice the temperature. There is no amount of temperature; it measures the condition or state of the stuff in the box. A number does not need to represent an amount of anything. In some sense, people know this, because total temperature sounds wrong to them.

We do not have thermodynamic equilibrium in the air, but we have a local equilibrium, so it is as if each point in the air is like a separate tiny box. But like the cream, if I divide the room up into tiny cubes, assign a temperature to each and add up the temperatures over the volumes, as you suggest, I would not have the sum of the temperatures at all. What would be the units of the total you calculate?

Nicholson: Uh, degrees-cm^3.

Thermos: Right. Whatever that is.

Now here is another question: For computing your average, why would you add up the temperatures of the cubes in linear form? After all, at the very least you should take account of the different density of air, water and human tissue, not to mention the pressure and humidity differences in the air itself. Why not weight every cube differently, since different materials store and conduct heat differently?

And why not square the temperatures, or take them to the fourth power? With the height example, the physical picture of the sum made it reasonable to add them up linearly, rather than adding, say, the square of the heights. But if you are averaging the kinetic energy of molecules, it makes sense to calculate the mean of the squares of speeds, because energy, which goes as the square of the speed, is physically additive, while speeds themselves are not. Or, since the Stefan-Boltzmann law tells us that equilibrium radiative energy goes as the fourth power of temperature, why not raise the temperatures to the fourth power before adding them up? On the other hand, if you want to compute the average resistance in a parallel circuit, you add up the reciprocals of the individual impedances. Maybe we should add up the reciprocals of temperature. There are an infinite number of possibilities.

The fact is, for computing temperature averages, there is no physical guide to tell us what kind of average to use. There is nothing to stop you from using the mean, but neither is there any physical reason to prefer it to the other ones I suggested. Consequently, if the mean is increasing, it doesn't tell us about anything physical. Some other average may be doing the opposite. Then where does that leave you? What is the trend in the average if you can't decide, on physical grounds, which average to use? With temperature, there is no basis on physical grounds to use a simple sum, some other sum or some other more complicated rule for averaging, because temperature is an intensive quantity.

Kutta: Shouldn't all averages do the same thing? I don't see how one average could trend downward while another can do the opposite with the same data.

Thermos: That's a good question. Let's find out.

[Thermos sets up a suitable placement of the four laboratory thermometers around the room. After a short time, he writes the measurements onto the board: 17°C, 19.9°C, 20.3°C and 22.6°C, respectively.]

Thermos: The 17° thermometer is the one at the window as before. The others vary a bit, but within the expectations of our discussion so far. Let's suppose now that we leave the thermometers where they are until next spring. Then we open the windows and a pleasant breeze blows in, mixing the air. All of the four thermometers read 20°C. Has the classroom warmed up?

Everyone in the half of the class by the window, please calculate the average temperatures using the linear sum divided by four. Everyone else, do the same, but use the kinetic energy rule: Sum the squares of the temperatures, divide by four, then take the square root. Do this for both cases, then find the difference between the two temperature values. [After much activity and conversation, the calculations are completed and the room settles down.] You folks on the window side, using the mean of the temperature rule, what did you get?

Student: We get +0.05°C, so the room is warmer in the spring.

Thermos: And those using the mean squares, what did you get?

Student: We get −0.05°C for the change, so the room is cooler in the spring!

Thermos: You see? If you have no physical reason to choose one average over another, then you are just making guesses. "Warming" or "cooling" of the classroom depends on your formula for the average and not on the actual measurements. But the averages are not physically meaningful. They are just two different statistics.

The same sort of dilemma exists for measuring climate change with a temperature statistic. This is not where the problems with average temperature end. We don't have time for that today. All I needed to show here was an example of averages that do not agree with each other in terms of magnitude or trend. Everything could have been different again if we had chosen four different places to measure. Can you imagine circumstances to be any better in the real atmosphere and oceans, which are subject to wider changes over the whole Earth, in which we are expected to find a smaller trend? What would happen if the places we measure temperatures are constantly changing? What does global warming mean, then, if it is only based on a trend in one of an infinite number of temperature statistics, which is only based on a convenient selection of a small number of shifting locations from a temperature field in three dimensions?

The solution to this is to look at what people are really concerned about in climate change. No one is really worried that the temperature might rise by a few degrees, even if it is only "on average." People experience changes like that every morning and every spring. This is no concern at all. What people are actually worried about are glaciers changing length, the sea level rising, the frequency and severity of storms changing and changes in rainfall patterns, to name a few. They are worried about change in all of the properties of the climate, most of which are physical aspects that have little to do with temperature, and certainly less to do with some official global temperature statistic.

Somehow many people have gotten it into their heads that the official global temperature statistic is the same thing as the "climate." Nothing could be further from the truth. The fact is that even if the official temperature statistic were nailed down and completely fixed, all these worrisome things could still happen! Alternatively, if the statistic did change, nothing guarantees that something deleterious needs to happen. There simply is a lot more going on in the atmosphere and oceans than temperature.

Furthermore, there are many intensive variables other than thermodynamics that would produce issues just as problematic. Why does no one talk about global relative humidity or global pressure? Is there a problem with global moistening or drying? Is there a global squeezing problem as global average pressure increases, if it is increasing? With a large enough increase in the number of molecules in the air, we could get crushed from the pressure.

What it really boils down to is that temperature is the first thermodynamic property that many learn, and for some, it is the last. The subtleties of the dynamics and thermodynamics are simply unpresentable, so the grand creations of modellers have no impact. Instead, the modellers must suffer the indignity of having their intellectual products cheapened by their portrayal as fancy thermometers. They are not thermometers and global climate isn't temperature.

Well look at the time. I've gone over yet again. I'll see you next week. And, uh, don't forget, this will all be on the exam.

[The students file out, quietly wondering what the exam will be like.]

Runaway Language and Metaphors That Feed the Doctrine

In this chapter, we contrasted the inappropriate expectations of certainty attached to climate models with the formidable theoretical, experimental and computational challenges of understanding climate in terms of basic science. We conclude by looking at some simplistic ideas about global climate that have come to eclipse the real science, even among some people who should know better. These ideas mislead people into a blissful certainty that feeds the Doctrine.

The first of these we already encountered above in the context of temperature. Professor Thermos would say that a global temperature statistic is mere metaphor and a dangerous idea to let loose in an unscientific society. It is too tempting to misconstrue global climate as being one-dimensional, in both the literary and mathematical sense (zero-dimensional in climate-model-speak). People are too ready to believe that such a statistic, because it is quoted with temperature units, is actually the temperature of something that actually bears on physical processes in the climate. The confusion is understandable, but it is confusion nonetheless.

What Professor Thermos did not touch on is why something so deeply and impossibly complex as climate prediction is confidently presented as something we know how to do and are already doing routinely. We are not exaggerating: There is now a proposal in legislation before the United States Senate to set up a "climate forecasting" service. Someone, somewhere, has got the idea into his or her head that we know how to do it. MEMO TO SENATE: *We don't.* It would be fine if this were just a new approach to fund research into long-term climate, but of course it is really a political way of attempting to make weather forecasting and climate forecasting seem to be on the same footing. They aren't.

The illusory notion of a global temperature allows us to substitute the relevant but difficult questions for an irrelevant but easy one, Is the world getting warmer? This in turn feeds into the unfortunate habit of doing science by metaphor.

Science by metaphor is always a risky business, and one misleading idea in this category has done more damage to peoples' understanding than any other. You have heard of the one we have in mind: *the greenhouse effect.* We know you don't want to think of it as a mere metaphor, and we are sympathetic, because it is not your fault. It is so deeply engrained in the popular

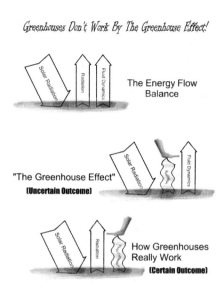

Figure 3.4

imagination that Chris has been told it's too late to complain about it. But complain we will, though we honestly wouldn't try if it were not such an integral part of the Doctrine.

The greenhouse metaphor is the secret mental model to which many retire when they become overwhelmed by the complexity and the definitive uncertainty of the climate science. The greenhouse metaphor is what made a referee once tell Chris that there is a law of thermodynamics that says putting carbon dioxide into the atmosphere leads to an increase in the (official) global temperature (statistic). Thermodynamics doesn't say that, but it sounds plausible within the greenhouse metaphor.

Ironically, greenhouses don't work by what has come to be called the greenhouse effect. This is the case of a simple analogy that grew to become bigger than the thing it helps to describe. Figure 3.4 illustrates how this came to be. The top picture shows the balance of energy flow into and out of the surface of the Earth, where we live.

Sunlight is absorbed by ground, and the heated ground acts like a radiator. When you add up the *net* amount of energy flow away from the surface by pure (infrared) radiation, it turns out to be roughly the same as that carried away by wind, air movements and evaporation. These are the two basic mechanisms for carrying away the energy: infrared radiation and fluid dynamics.

The greenhouse mechanism has to do with the balance of the flow in and out. It is a simple idea, and there is nothing at all wrong with it. If there is more coming in than going out, the energy is piling up. If there is less coming in than going out, the energy is being drained. When no piling up is taking place, everything is balanced. The fluid motion cooling and infrared radiation cooling each carry away about half of the energy brought in by sunlight.

A greenhouse acts like the picture on the bottom. Someone comes and shuts off the fluid dynamical energy drain from the surface by putting something up, such as glass or plastic, that the inbound solar radiation can pass through but air cannot. It doesn't matter what the material is, the effect is quite pronounced. The explanation is both theoretically and experimentally certain.

After the fluid flow is shut off, the energy flow is no longer in balance. The incoming energy cannot escape fluid dynamically. The balance can only be restored by radiation. We know how to solve the equation of transfer that governs radiation, and we know the solutions call for the flow to increase with temperature. So that's it. The temperature will go up, restoring the balance, and there is absolutely no doubt about it. It is certain.

It is also irrelevant. "The greenhouse effect" doesn't work that way. It works according to the middle picture. Infrared absorbing gases, increased within the atmosphere, restrict the radiative drain of energy. Again, there is too much radiation coming in for what is going out to be in balance. The temperature then goes up at the surface, causing the radiation drain to increase, leading to a happy and simple balance. Ergo warming, right? Wrong.

Unlike the greenhouse, we still have the untouched, equally large fluid dynamics arrow to contend with on the right side of the middle picture. As we have emphasized throughout this chapter, this is largely turbulent, and we don't know what it would do. We can't solve the governing equations. Recall that in the case of turbulence we can't even forecast from first principles the average flow in a simple pipe.

We do know that to increase the drain of energy from the surface fluid dynamically, we do not need to increase the temperature. Fluid flow rates are governed by more variables than temperature, and these typically appear as gradients. So we don't need to look for an increase in *any* variable to lead to increased energy drain through fluid dynamical processes.

This is already more than enough to make the point that the greenhouse effect won't likely work like a greenhouse, but there's more. Fluid dynamics also affects the other two arrows at the surface. It would be so much simpler if it didn't, but it does. The fluids carry dust and aerosols into the atmosphere. These can significantly affect the infrared radiation problem. With the thermodynamics, they also carry water vapour into and out of the atmosphere. Not only do these significantly affect the infrared radiation problem, but they determine how many clouds and aerosols throw the sunlight out of the atmosphere before it even gets to the surface. That means the big arrow on the left can be reduced or increased.

What will it really do? It is actually very simple in the end. If you do not know how far the blinds are pulled or how high the air conditioning is set at, you cannot say much about your greenhouse. That is the case in the middle picture. We have blinds (clouds and aerosols) and air conditioning (fluid motions and evaporation from the surface) and they are periodically adjusted and we cannot tell what they will do. Could it be more unpredictable?

However, by calling the middle picture the greenhouse effect, the certainty of real greenhouses gets grafted onto the global climate problem. So our official global temperature statistic has to go up because greenhouses are sure things. That is the metaphor to which people's minds retire. It is unscientific nonsense. But it props up the Doctrine.

Everyone says the models support the Doctrine. But they really are not convincing in themselves, because they are big, confusing and uncertain. With the greenhouse dodge, we can say that the complex modelling issues are very interesting, but we really don't need them because we know it is really simpler than all that.

Incidentally, the greenhouse idea has become so entrenched that you can read textbooks that say greenhouses work by the middle picture. They usually envision some sort of glass with special properties varying with the type of light. But they usually ignore the fluid dynamical arrow. There is no mystery here. The power of metaphorical thinking just sucked people in.

Why was such thinking started? You have to bend over backwards to explain things in an unscientific age. So someone once tried to explain flows, in and out, by a greenhouse metaphor. This was so compelling that, ironically, greenhouses came to be explained by atmospheric radiation flows. It's a tradition that everyone trained in atmospheric radiative transfer laughs about, because they know the truth about the missing fluid dynamics.

However, in the 21st century, it doesn't seem like such a laughing matter, as nearly everyone takes the idiosyncratic language that describes the middle picture as the greenhouse effect at its face value. Reporters have been known to visit greenhouses when filming stories about "global warming." In one television news story, a science reporter put a little model village under a clear plastic dome. The hot television lights heated it in exactly the same way as sunlight would. The dome set up the circumstances of the bottom picture in Figure 3.4. With great ceremony, the rising temperature on the thermometer under the dome was observed in prime time as proof that global warming would be coming to the viewers at home. What nonsense!

So-called greenhouse gases have absolutely nothing to do with greenhouses. We will call them infrared-absorbing gases (or infrared gases) here. The most important of them, radiatively speaking, isn't carbon dioxide, it's water vapour! Water is more important to the radiative transfer of energy than all of the other infrared gases combined, and there are about a half-dozen usual suspects. However, too often, water vapour will not be on the list of "greenhouse gases" even within some professional discussions. Moreover, its obvious link to the fluid motions, clouds and aerosols will often be omitted when it is mentioned.

Because of that link to the vast uncertainty that fluid dynamics and the rest of the fundamental climate problem represents, leaving water vapour off a list of infrared gases amounts to leaving off the uncertainty of the climate problem.

An auxiliary issue that helps to obscure the significance of water vapour is to talk about the "carbon cycle" instead of the carbon dioxide cycle. The cycle refers to the creation and destruction of the infrared-absorbing gas carbon dioxide in the atmosphere. The term "carbon cycle" is a stylistic term, but carbon itself has no role in the greenhouse effect. Carbon is found in pencil "leads" and diamond rings, rocks and plastics that never degrade. What can make a carbon atom relevant is if it is connected to two oxygen atoms. Carbon cannot do it alone. But if carbon is "bad" with oxygen, hydrogen is even "worse" in the form of water.

One famous Canadian environmentalist, Dr. David Suzuki, explained global warming in the *London Free Press* (May 12, 1990) by saying that automobiles get hot on a sunny day because the "carbon" in the glass caused the warming. This is a sort of second-order nonsense, combining the misleading "greenhouse" metaphor with the misleading "carbon cycle" jargon into one error, squared. Cars warm because they are *real* greenhouses, as in the bottom picture in Figure 3.4. You don't need any carbon to be mixed in with the glass to do the job.

But you can see why non-experts get the wrong ideas by using the wrong words and metaphors. At some point, people start to think that it is the carbon atom that is bad, and that throwing a pencil into the air and catching it has something to do with the carbon dioxide cycle. In the process, water—which has no carbon—gets forgotten. When water is forgotten as an infrared-absorbing gas, the whole unsolved climate problem fades from sight and the Doctrine grows.

Most public discussion of global warming in the past few years has been built on incoherent clichés and misleading metaphors. Here's an example from the Environment Canada web site.[2] Under the heading THE EARTH IS A GREENHOUSE we read: "As you know, greenhouses use glass to keep the heat in. And just as the glass in a greenhouse holds the sun's warmth inside, so the atmosphere traps the sun's heat near the Earth's surface." This is part of an "official" federal government Web site published to inform Canadians about the climate change issue. Official or not, it is bunk that only serves to confuse the public and reinforce the Doctrine of Certainty.

Of course such problems are endemic. A global temperature statistic is endlessly discussed despite the fact that it is not a temperature. It has little to do with rising sea levels or the length of glaciers, yet its every movement is said to control these things. Water vapour is king among infrared gases, yet is rarely mentioned, even though the behaviour of other infrared gases cannot be understood unless you can figure out everything due to water vapour first. Carbon is not carbon dioxide. Simple, but it needs repetition.

There are no controlled experiments in climate, and above all there is no theory for climate. This pretty much crosses everything off the list of things we could do to treat climate prediction definitively. All that is left is to stare at tea leaves, look at models and analyze observational data using statistical methods. We leave the first to Official Science. We will do the latter two in the next chapters.

---

[1] Sir James Lighthill. Proceedings of the Royal Society A 407, 1986, 35–50.

[2] www.climatechange.gc.ca/english/workroom/students/ greenhouse.shtml, January 24, 2002.

4 T-Rex Devours the Planet

We explained in the last chapter that our challenge is to think scientifically about a scale of nature we do not experience directly, namely, the climate. Measurement and theory will likely involve some kind of averaging over our everyday human-scale observations. But the ascent from the theory at the atomic scale to that of the middle realm taught us a key lesson: Not just any averaging will do. We gave an example of fluids, in which averaging done plausibly but incorrectly yields equations that do not describe actual fluids and contain averaged variables that do not correspond to physical reality. Not all averages give us paths up the mountain: Some lead to dead ends and some lead us over sheer cliffs.

In this chapter, we are going to talk about the whole business of averaging temperature. There is an intuitive sense that studying climate requires temperatures to be averaged. Based on that intuition, people are going ahead and making all sorts of averages, in all sorts of ways, with no theoretical basis for any of them. And, not surprisingly, they do not necessarily agree with each other. Worse, we are not quite sure what any of them mean.

In the last chapter, Professor Calorie J. Thermos raised a basic issue about temperature that is rarely discussed in connection with climate. Averages of temperature are physically problematic. However, they are so unquestioningly accepted that we must revisit this issue at length. There are great debates about the process of creating the averages and there is a naive faith that all averages "properly" produced must be the same. We have already seen that this is not true in Professor Thermos's classroom, but this story is so deeply ingrained it will take more than one telling.

Panic in the Streets

April 2002 in Ontario began with hot, summer-like conditions. The news offered up man-on-the-street interviews with leading questions about global warming. But this only lasted a couple of days. The weather suddenly became colder—20°C to 30°C colder. There was snow on the roofs in late April. There was snow in the first weeks of May and ice pellets in the middle of the month. These are not "average" conditions here by any means.

While many were looking forward to warm weather, for us it was a relief. Polls showed belief in global warming fell during this time. For a few quiet days, there were no "global warming" scare stories. But they returned as soon as it was warm enough for newsroom editors to have to turn the air conditioning back on in their cars. Curiously, there were no reporters sent to investigate whether the spring snow signalled a coming ice age. At one time there would have been, but there is no longer equality among the hysterias.

As Chris drove his car in the middle of April, granular snow rattled off the windshield. But his car thermometer measured the external air temperature at 6°C (42°F), which is well above freezing. How can there be snow when it is so far above freezing? It's only a paradox if you think there is only one temperature in the world, or if you think nothing is going on except hot and cold.

But people who talk about climate change want to focus on temperature, to the exclusion of just about everything else. They talk of some temperature number instead of the temperature field. The irony is that the only world in which you could talk about climate in terms of a single measure is one in which there could never be climate change!

To restate one of Professor Thermos's points from Chapter 3, temperature is not a single value in any physical system except one in *thermodynamic equilibrium*—a state of the system where all the processes that are going to happen have happened; everything is played out and nothing changes anymore. The Earth is not in that state at all. If it were, we couldn't have climate change at all and there would be nothing to talk about in this book.

So on Earth, there is no single value for temperature. That is why ice pellets could be formed even when the air temperature was 6°C. They were formed where it was colder and transported to where it was warm, just outside of Chris's windshield.

There aren't just a few temperatures either. There are more temperatures than there is room for thermometers. At every point, there is a temperature for the material in that spot. It gets even more complicated if we talk about the atmospheric radiation field, so we won't. There are infinitely many temperatures to be measured because there are infinitely many temperatures around us. There are infinitely many in the mountains and in the clouds and at the bottom of the ocean. There are temperatures everywhere. They are always changing, and any particular place does not need to have the same temperature as any other particular place, even if no two differ terribly much from each other. Temperature is a *field* and not a single value. It is not the only physically relevant field. It really isn't that special. But temperature is the only quantity that most people talk about on the subject of climate, so we have to talk about it as if were special.

To specify the temperature field, you need to know what all the temperatures are, at every place on Earth (in the ground, sky, and water) at a particular instant. That is a lot of information to hold, but it is still not enough to tell you what the temperature field will be in 10 seconds or 10 months. The temperature field does not contain the information necessary to tell you what it will be at the next instant.

To do that, you need to solve the whole physics problem discussed in Chapter 3, which involves other physical properties than just the temperature field, such as the pressure field, for example. As we saw, that took us up against the great unsolved problem of turbulence and related hard-nut problems of modern science. No one knows how to deal with them.

Nonetheless, suppose that we could come up with the full temperature field at every instant over a long interval of time. Some people would like to treat this collection of temperature fields as random, even though they are not really, and nearly everyone would like to replace them with a single number for each instant, even though the physics doesn't tell us how to do that with temperature.

Recall that Professor Thermos explained that temperature is an *intensive* thermodynamic variable. So no one rule for making an average out of the field is preferred over another. Furthermore, he showed us that some types of averages go up while others go down when applied to the same data. Different averages do not all say the same thing. You need some specific physical reason to prefer one average over another—and there isn't such a reason for fields of intensive variables. This is Nature's way of telling us that the temperature field should be left as a field and that no single number is going to be able to stand in for it.

But people often don't listen to Nature's little hints. They go ahead and pick one way of doing averages anyway. It has no physical basis. It is just a tradition, and it seems like it is special, but it is not. The single number people use in replacing the field at some instant is not called *an* average, as it really is; it is called *the* average, as if there is no other choice because of some mysterious law of statistics.

We will see in what follows that different averages from climate data produce different behaviours. Professor Thermos would likely yawn at this, but it has caused great consternation. The Doctrine discounts alternatives and looks only at one particular average. It is so much more compelling if the climate is just doing one simple thing rather than an infinity of ambiguous, complex and inscrutable things. It is so much more monstrous to have all eyes locked in fear on one towering beast with one overpowering message.

When we think about reducing the dimension of the problem, we quickly run up against unanswerable questions. Suppose we ignore all of

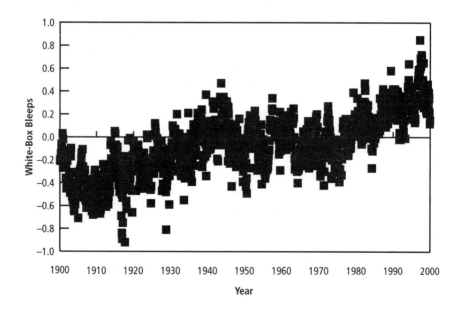

Figure 4.1. T-Rex, January 1900 to December 2000.

the points in the atmosphere and oceans except those right at the surface. This is not the complete temperature field, but suppose we are only interested in the surface values. We still need an infinite number of values to determine the temperature field at the surface, even if we intend to collapse that further to one number.

But now suppose we don't have all of the values at the surface. Suppose we are missing an infinite number of values at any one instant except for a few thousand or so. And one more thing: Suppose we are missing an infinite number of instants when the temperature field is measured, and... well, let's not worry too much about the rest of that for now.

Even given the pale shadow of what is left over, there are still an infinite number of averaging rules that can be applied to the data we do have. Out of all those choices, a particular class of rules has emerged as the custom. There is still some considerable wiggle room, which we will discuss later, but Figure 4.1 is the result of applying all customary traditions and rituals to the remaining data from the temperature field to get a single number over time. Many people really like it. It is everything they always dreamed of.

The Web site where we got these numbers calls it the "Global Temperature."[1] Of course it isn't the global temperature. There is no such thing. It is just one of an infinity of possible numbers that could be distilled from the infinity of numbers making up the temperature field at any instant. This number is properly called a statistic, if it is done consistently.

The plot shows a sequence of monthly values across the 20th century. Though the numbers refer to periods back to the beginning of the last century (and there are more back as far as 1856), it is only fairly recently that people have been assembling and graphing this sort of statistic. Consequently there is still very little understanding of how it is constructed and what it means. But it is a wonder to behold, and people (even experts) assume they know what it means, so it now occupies centre stage in the climate change horror flick.

In some quarters, Figure 4.1 is so identified with temperature, and so central to conversations about climate, that it is just called T for temperature. In those quarters, people like it so much that you are just expected to know they mean the numbers in Figure 4.1 when they say T. And for the Doctrine, T is king: T-Rex. It devours all other scientific sensibilities. No thinking about any other physical property of nature seems to matter. For the Doctrine, T—augmented by omens, portents and signs—is all that exists.

Among those who consider *T* king of the climate, there exists an unstated but widely held belief that it ought to be flat. If it rises, as it does in Figure 4.1, this is thought to be cause for alarm. It is not clear why T-Rex ought to be supine, nor what an upward slope really means. But because it is called the "Global Temperature," an upward movement gets called "Global Warming." And the Doctrine says Global Warming is bad. Unravelling these ideas will take the next four chapters. In this one, we start by telling you what T-Rex really is and isn't.

Think of Figure 4.1 in terms of a bad B-grade horror movie. Some mad scientist sets out to reconstruct a dinosaur. But he cobbles together something else instead. Out of real bones, he gets some weird and fearsome monster that never actually existed. He produces a scary picture of his work that frightens many people. Massive panic ensues, with people running terrified through the city, causing mayhem and destruction. So the UN calls in a new superhero, the great Green Kyoto, to subdue the beast and build a cage powerful enough to hold it. Hooray! We're saved! Roll credits.

Okay, no one would believe such a silly tale. Why would anyone panic about a scary picture? Especially if it's not clear what the thing is or whether it exists at all.

On second thought, maybe people *would* believe such a story. After all, this is pretty close to what has happened over Figure 4.1. People are panicking over a picture. They assume that it is a picture of something that has physical reality. They presume that it shows something terrifying is happening to "the" temperature. In their panic, a plan has been hatched to save them, keeping "the" temperature contained within some "safe" bounds.

But this isn't the temperature of the Earth. It isn't the temperature of anything at all. It's just a statistic that has become a sociopolitical monster. Professor Thermos would say it is not hard to see that it is quite disconnected from the physical world. He would say there could be a great climate disaster without any movement (upward or downward) in Figure 4.1. Alternatively, Figure 4.1 could jump off the page without anything important changing in the world around us. Trying to frame the climate change issue in terms of up or down movements in Figure 4.1 is an example of the sort of lazy metaphorical thinking that serves to cloud the real issues and prevent deepening of our understanding. As such, it serves the purpose of the Doctrine quite well.

We will start by briefly telling you some details about how T-Rex is put together. We will also tell you about the debate over whether it is reliable. Some of the arguments take for granted that it is a real physical object instead of many ambiguously related mathematical ones. From our point of view, this debate is like the old chestnut of how many angels can dance on the head of a pin—except that no one actually debated that question.

The debate itself—seeking the best way to determine something without physical meaning—is ultimately fruitless. Figure 4.1 is not what it claims to be. It is not the "Global Temperature." We are not entirely sure what it is. But whatever it is, it haunts the nightmares of many people as T-Rex, ferally devouring physical sensibilities that people would normally have.

The Bones of T-Rex

Many readers of this book will know almost nothing about what sorts of numbers go into T-Rex or their origin. But they are vaguely aware that they ultimately come from thermometers, and people are comfortable with thermometers. Beyond that, they care little about the details: They figure it's the "global" temperature.

Maybe they don't actually think there is a global thermometer at the global temperature station, but people talk about a particular type of temperature average all the time. They have a sense of the "regular" temperature on a given day. TV weather forecasters call them "normals," as if to suggest it is normal for the temperature to have that value.

They are usually thinking of a mean when they talk of normals. The mean is just the sum divided by the quantity of data. In what follows here, "the average" or "the normal" will always refer to a mean. But we will still need to say specifically what folks mean by the mean.

Statisticians call the difference between the normal and the observed temperature the "anomaly," as if anything other than the "normal" is anomalous. This could confuse a person who is not careful. A normal is just a mean value at one place over a particular period of time. But you don't get all of the local temperatures, just ones measured at specific instants.

Someone needs to choose over what period of time to collect the values that go into the mean. There is no physical rule that can be followed. The interval will be arbitrary and the average need not be the same over all intervals of time. Choose a different period and the "normal" will change, even if the laws of physics don't.

Suppose we choose an extremely short period over which to do an average. Suppose instead of years we choose only two days. One day in June at 6 p.m. it might be 16°C, and the next day 20°C. If you define the normal as the 6 p.m. average across those two days, you would conclude the first was "colder than normal" and the second was "warmer than normal."

If instead you look at those two days in terms of an average over the previous 30 years, it may turn out that both days were cooler than normal. But there is no reason to average over 30 years. You can average temperatures over the previous 300 years. Over that span, you might find both days were warmer than normal. The temperatures stay fixed; it is just what we compare them to that says whether they are above or below normal.

There is a convention to use a period of exactly 30 years to define a normal. Why 30 years and not 20, 42 or 300? There is no physical reason except tradition. The tradition might have called for a longer period, but we do not have long enough weather records in most places to have the choice. But that is not a scientific basis for choosing one over the other; it is just a tradition originating from consistency and convenience. If we found something else to be more convenient, we might get different answers as to whether our two days were above or below normal.

There is no reason to say the average over 30 years is the "normal," while the average over 300 or 300,000 years is not. There is no "normal" temperature, if by that we mean some temperature at which the thermometer should permanently sit. Still, weather forecasters like to tell us about normals. It might be that the temperature today is above or below the numerical average of a particular 30-year interval. If people find that interesting, so be it. But there is a problem with using this kind of language.

By adopting terms like "normal" and "anomaly," we have snuck a large, unstated set of expectations about the behaviour of climate into the discussion about temperature. These expectations do not proceed from a scientific theory. There is no theory of climate. They proceed instead from a combination of convenience and tradition. There is nothing normal about a "normal," and nothing anomalous about an "anomaly." In fact it would be highly abnormal if there were no "anomalies" and the temperature were always "normal." These terms are just jargon for numbers that describe what is happening today compared to what happened in the past. The statisticians who use this jargon just don't use the words according to normal English usage, and you should not interpret them as such.

Bleeps

T-Rex is built up from these so-called anomalies and normals, which in turn are derived from temperature readings on thermometers located in some white boxes with holes in the sides, mounted on stands and scattered here and there around the world, mostly in the U.S. and Europe.

We have said already that T-Rex is a statistic and not a temperature of anything. That is a simple enough idea, but it remains a mathematical consequence of the procedure to get the mean that it still has units of temperature. You can convert the statistic from Celsius to Fahrenheit, for example.

However, if we repeatedly say that T-Rex increases or decreases by, say, 1°C , how long will you be able to resist thinking about changes in terms of the temperature of the air surrounding you? How long will you be able to resist imagining that there is some global thermometer station that measures the temperature of the whole Earth?

You would not be the first if you find yourself thinking that way. Many very smart people have been unable to resist the impulse. "Global temperature" cannot enter the front door of scientific discourse because it makes no physical sense, but it has snuck in a semantic back door when no one was looking.

So we have a plan to help. We have developed a new term to describe Celsius units of change for the statistic shown in Figure 4.1. The special unit will indicate that we are not talking about the temperature of a physical thing in nature but values of a statistic instead.

We call it the "Bleep," abbreviated to "B." By measuring change in Bs instead of Cs, we will remind you that you are using a statistic and not a temperature while permitting it to have its units. You can always say to yourself that a Bleep would be like one degree Celsius, but then the concept would have to enter your mind through the front door.

The Bones of Contention

Temperatures as read from the thermometers in white boxes (also called stations) are not used directly in producing the data for Figure 4.1. Instead they are shifted by subtracting the local "normal." These shifted temperatures, or local anomalies, go into a weighted mean. Stations in areas with lots of nearby stations are each weighted less, as are stations further toward the poles. This is called **gridding**, and it is intended to scale station readings up or down based on the amount of the Earth's surface each one is taken to represent.

This is a model assumption. It says that stations that are far from others better approximate the temperature field than stations that are close together at the surface. It is not unreasonable, but it does underline that we are not working from the outset with the true surface temperature field but only a shadow of the real thing.

It also means that the average is based on weighting by surface area, so the gridded temperatures aren't actual measurements. There is no reason to weight the thermometer sites this way physically, of course. We could as well weight it by some assumptions about, say, volume or mass or nothing at all, which would each lead to a slightly different mean. This would be a bothersome idea for someone who believes that there is only one "average," which includes a great many people, unfortunately.

Since the physical world does not determine which of the many physically natural averages to use, as it could for non-intensive quantities, averages over temperature have no particular physical significance. Instead of ruling out all averages, oddly, that puts all possible averages into contention. They are not physical variables, but each one is a statistic. No two statistics need to behave the same as the other, as we have already seen. We will find this is so for different statistics from the full temperature field in the atmosphere, although people toil and struggle to prove that they ought to be alike, because they think wrongly that their pet statistics are genuine physical variables.

In the following, we will find that even competing attempts to measure the very same statistic do not agree because of the slightest differences in how numbers are harvested into slightly different averaging processes. Moreover, they do not just differ in magnitude, they also have slightly different trends. They present themselves in this way because they are different statistics, even though they are intended not to be. This difference in trends is widely ignored.

When a statistic from the full temperature field is computed completely differently than those others, as it is in the satellite and balloon record, substantial differences in behaviour emerge. While this should present no surprise, the different behaviour leads to disputes about which one is right, even though there is absolutely no reason why they must agree. They are all just different statistics.

A statistic is lower on the pecking order than a physical variable because it doesn't fit into any physical theories or necessarily have any significance at all. When a statistic is formalized, it is elevated to the status of an "index." But even lowly statistics or indexes have requirements for consistency that we shall see T-Rex fails to meet.

This means we are arguing about things about which there is no problem, ignoring things that are significant, and toiling over something that is not even a single proper statistic, all because there is this flawed notion that a single temperature can represent an open physical system that is not in any sort of global thermodynamic equilibrium. That monstrous confusion is what T-Rex is about.

Cities

As we mentioned, most of the boxes are in the U.S. and Europe, some are in Canada and Australia and other countries, and a very few are in developing regions like Africa and South America. As far as the spread of boxes goes, you will not find many out in forests or up in mountains or other remote places. They are mostly in towns and cities, especially at airports. This is one of the points of debate over the data in diagrams like Figure 4.1.

Critics argue that a lot of the "anomalies" are merely registering the fact that cities are getting bigger and hotter over time. In a way, people who produce T-Rex (and the Big Panel) concede this possibility, for they claim they adjust their data for "urban heat biases."

While, at this stage, readers of this book should wonder why this matters, you should bear in mind that the proponents and critics alike think there is a "true" global temperature that these averages supposedly approximate. Thus their debate turns on whether there are too many data from the warmer cities. From this mistaken perspective, this is crucial, because the entire case for "global warming" hangs on what this "true" temperature is actually doing, even though no true global temperature actually exists.

So they aren't "biases" in the sense that the temperature field does not have cities in it. Wherever you put your thermometers is just the piece of the actual temperature field that you sample. You will never get the whole thing with the few thousand thermometers used, no matter where you put them. Furthermore, T-Rex teams have empirical formulas that take as inputs the raw thermometer data from, say, Atlanta, and spit out estimates of the local air temperature would have been if Atlanta were not there. So they *do* think of some kind of world with no cities when they make up T-Rex.

Nonetheless, they have concocted some rules that translate the actual data into a form that supposedly represents a "pure" climate signal. The techniques look for things like correlations between local temperatures and population growth, or conspicuous differences between close pairs of stations where one is in a city and the other is outside it. (It should be noted in passing that critics have shown you can have urban heating without population growth, and the pairwise-comparison technique is only useful if you have a lot of stations to work with, which is rarely the case outside the U.S.)

Regardless of whether these correction rules make much sense, what it shows is that T-Rex is composed of numbers that are outputs of a lot of models, not just a direct reading of some thermometers.

The rules that produce such numbers are models as surely as those discussed in Chapter 3, complete with all the potential for misunderstanding and error. Not only is the temperature field created from a model that changes the original data, but many of the properties of the model have an ad hoc nature hidden from the casual observer as surely as the ad hoc aspects of any other models. The basic data for T-Rex come from white boxes, but then they go through statistical black boxes on the way toward Figure 4.1.

We do not have controlled experiments. We do not have a theory. Now the meaning of the "basic observations" themselves are in dispute, and they are all subjected to ad hoc adjustments.

Oceans

A glance at a globe will remind you that we live on a *blue* planet. Most of the Earth (about three quarters) is covered in water. And there are no little white boxes on the water. There are some buoys here and there with thermometers on them, mostly in the tropical Pacific, but not enough to provide much coverage. This creates a big problem for claiming T-Rex is a "global" anything.

There are two teams that produce Figure 4.1-type diagrams, and they deal with the lack of ocean air temperatures differently. The team at the Goddard Institute of Space Science at NASA just ignores the problem and publishes a statistic based only on data from boxes on land. That implies a particular averaging rule: Take the air temperatures over the ocean and give them a zero weight.

The other team that produces T-Rex numbers is at the Hadley Centre in the UK. Their data went into Figure 4.1. They combine the air temperatures on land with some *water* temperatures from ocean-going ships. According to the gridding rules, measurements far from neighbours will be weighted more. So while the Goddard model is to weight the ocean low, this picture weights it high. Once again we have two rules and two different averages.

Since early in the century, commercial ships have been taking water temperatures on their voyages and supplying the data to the UK Meteorological Office. They used to haul a bucket of water on to the deck and use a hand thermometer. Around World War II, however, they switched to recording the temperature of the water going into the engine cooling system. Back in the 1980s, people discovered that the switch from bucket to engine-cooling water data had caused an upward creep in the average readings over time, but there was no way to estimate how big it was or how to remove it.

More recently, a group of scientists[2] compiled water and air temperature data from a network of buoys, ships and weather satellites for a large region of the tropical Pacific. Since 1979 (the period they studied), it turns out air temperatures above the ocean surface and water temperatures just below the surface have been diverging.

The buoy network data are especially useful since they measure temperatures at one metre below the surface and three metres above it in the same location. In all comparisons of water and air temperature at the same location, they found that the ocean is warming relative to the air. Moreover, three of the four temperature data sets (satellite, balloon and buoys) indicate the air just above the ocean surface has been getting colder throughout the tropics, even while the water has been getting slightly warmer. Yet it is the water temperatures that get put into Figure 4.1. Monitoring water temperature turns out to be not such a good way to estimate changes in nearby air temperatures: The two materials can (and do) warm or cool in opposite directions.

Professor Thermos would probably make some specific remark about the relatively high heat capacity of water and wonder generally why anyone would be surprised that the temperatures would be different. He would point to the differences over the 20th century between the Hadley Centre and the Goddard Institute statistics (Figure 4.2) rather than their

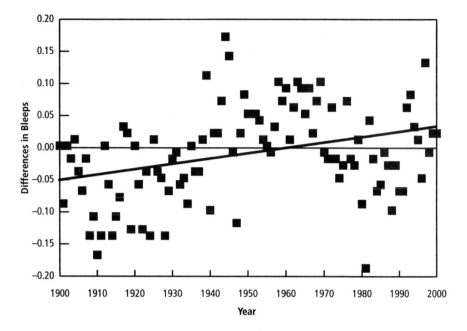

Figure 4.2. Differences between two different "global" temperature statistics. The Hadley and Goddard Institute statistics were scaled to average zero over 1961 to 1990, and the dots are the differences between the two resulting series.

similarities. He would say that this is an example where two different statistics distilled from the temperature field also behave differently. Where is the surprise?

Note the differences are not getting smaller over time. Despite the similarities in the rules used and the raw data, they are ultimately different statistics from each other, even though they belong to a similar class of averages. And the magnitudes of the differences (plus or minus about 0.1 Bleeps per year) are large compared to the variations in means that people look for as "global" warming.

Satellites

The thermometers in white boxes and water buckets measure temperatures near the surface. Weather satellites measure temperatures in the troposphere, which is just another (larger) part of the global temperature field. None of the statistics sample from the whole temperature field, but there is no reason to believe that one piece is more important than the other. In each, the weights of large expanses of the actual temperature field are zero.

The satellites observe the intensity of microwave radiation from oxygen in the air emerging from above the atmosphere. This radiation tells us the temperatures of the whole atmosphere. Knowing the details of the radiation, they figure out what the temperature field of the atmosphere below about 15 km (the troposphere) is. Figuring out the temperatures that produce emerging non-equilibrium radiation is known as an inverse problem. These are famously difficult, but we can do this one very well.

They take measurements from the equivalent of about 100,000 places from the temperature field each day. Each point would be about 40 km from its neighbour, if you assume that they are uniformly distributed over a volume equivalent to the troposphere. However, as the troposphere is only about 15 km thick, if you distribute them uniformly over a surface area equivalent to the Earth's, the neighbors are only two to three kilometres apart. This is very fine coverage compared to the surface network, which would sample only about 15,000 places at its peak historical coverage on the surface. Its points would be about 180 km between neighbors. Of course they are both coarse coverage compared to the infinity of points in the actual temperature field.

With these reconstructed temperatures, a mean rule produces an average temperature too. Even though it too replaces the very same temperature field with a single number as in Figure 4.1, it is an entirely different averaging process. The satellite temperature statistic that people calculate from the weather satellite data covers the whole troposphere. Their result is shown in Figure 4.3.[3] As you can see, in contrast to Figure 4.1, there is no upward trend.

In Professor Thermos's demonstration in Chapter 3, the trend behaviour of an average is a property of the average as much as the data. One average rule can be going up when another isn't. Accordingly, this presents no great surprise for an out-of-equilibrium physical system.

Yet here is where the misguided ideas about averages appear. Even though the field average from satellites is just as legitimate or as illegitimate a statistic as T-Rex, Figure 4.3 was not included in the SPAM. T-Rex, on the other hand, has its claws all over the report. A balanced view of such statistical indexes would be to discuss everything that was available.

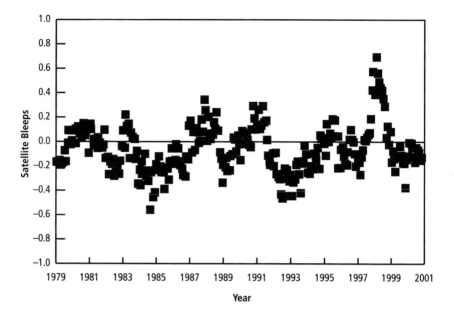

Figure 4.3. Tropospheric average temperatures as measured by weather satellites, 1979 to 2001.

But many people think the universe is not big enough for more than one average. There are enemies of T-Rex who think that the satellite average is *the* true one and the surface average is so much crap. The Knights of the White Boxes respond that the satellite averages are very silly and no one should pay any attention to them. The Defenders of the Satellites went before the Grand Council of the American Meteorological Society and were given golden medallions for their work. They then rode forth and smote the Knights of the White Boxes. Then all the people cried out in confusion and the High Priests of the National Academy of Sciences inquired of the oracle. It issued a report declaring that everyone is right and we should all just get along.

The National Academy was right that everyone was right. There is nothing inconsistent in having two different extracts from a single collection of data, both characterized as "the" average, which do different things. It would be remarkable if they were the same.

But what do they mean if they are both "correct"? We do not know what Figures 4.1 and 4.3 would show in a world in which no one ever used fossil fuels. Any particular average made of data from the climate can show just about anything without its being evidence of something unnatural. We will talk about this more in Chapter 6.

Another point of the controversy is what Figures 4.1 and 4.3 say about climate models. Standard runs from GCMs suggest that if carbon dioxide is the culprit behind global warming, then the average from the satellites, which is of the whole troposphere, ought to be going up as fast or faster than T-Rex from the ground stations.

The Big Panel gets around this difficulty with some clever statistics. They quote averaged rates implying that the two statistics do behave in the same general way even though they are manifestly different. They make the argument using another statistic similar to that from Figure 4.3, which is derived from some weather balloon observations back in the 1950s.

Back in 1976–77 there was an abrupt event which today is called the Pacific Climate Shift.[4] It seems to have been caused by simultaneous changes in the Pacific circulation and two atmospheric cycles: the Pacific Decadal Oscillation and the North Atlantic Oscillation. This created a step-like increase in temperature at many spots around the world, especially around the North Pacific, as well as changes in fish stocks and other climatic indicators. At many of these places, the trend before and after was effectively zero.

The Pacific Climate Shift is not part of the "trend," it is an interruption to the trend. But by running a trend line through the whole sample from 1958 onwards the slope is made to look about +0.1 B/decade. The PUN says that the satellite data look like the balloon data, ergo they show an increasing trend of +0.1 B/decade too. Hence the satellite data shows "warming," just like the surface data. However, one is a step and the other is a slope—very different.

But if you are trying to say that the whole Earth is doing just one thing, as you pretty well must if you are talking about *global* warming, then all of the statistics had better be singing the same song, and you run a linear trend line through them even if there isn't one. This is of course obfuscation. Anyone who looked at Table 2.3 in the Big Panel assessment report could figure out the game, but who looks at the full scientific report? But that is how the Big Panel handled the problem of the satellite data and preserved the idea that all statistics are the same.

Spatial Coverage

If you can only measure the temperature field at a few particular locations, the spaces where you don't have a thermometer are simply not measured. You just don't know what goes on in between them, unless you make the perfectly reasonable assumption that not too much goes on in between. But you can't say for sure. You don't get something for nothing.

Unless you want thermometers everywhere, that assumption has to be made at some level. However, it can go too far too. Some critics of T-Rex argue that there are too few thermometers to provide meaningful coverage of the Earth's surface, and the ones that are available tend to get moved or closed so often that it compromises the continuity of the statistic.

Spatial coverage is a slippery aspect of temperature measurement, because the physics doesn't support it. Temperatures have no associated volume or area, but for an index like Figure 4.1, we want a thermometer to represent temperature for a volume or area anyway. Essentially we are looking for a weird, non-thermodynamic entity in units of degree-metres-squared or degree-metres-cubed, because temperatures are not per-unit anything.

The atmospheric temperature field exists in all three space dimensions. So, ideally, the field should be sampled over all 8,000,000,000,000,000,000 cubic metres, if you include the troposphere up to about 15 km. Since we have the satellites measuring the free atmos-phere, we can focus only on the surface of the Earth, which is about 500,000,000,000,000 square metres. If we put one thermometer at the surface for each square metre, we will always be bumping into them or running over them with the lawn mower. Thermometers placed one metre apart are probably unnecessary for the most part. So we have a high-end limit of 500,000,000,000,000 thermometers in the network versus the low-end limit of one thermometer. Clearly a practical number must be selected from between these two values. What is reasonable? Quite apart from the practical constraints imposed by the happenstance of where the white boxes are located, people have looked at the question of where they ought to be and how many we could hope to make do with.

One thermometer outdoors pretty much tells you how to dress on a cold winter's day for as far as you would likely walk before it starts to get dark and the local temperature changes anyway. That is, say, about 10 kilometres at most. That corresponds to an area of 100 square kilometres, or 100,000,000 square metres, and would imply 5,000,000 thermometers in the network. This number might be fine for some cases, while in some other cases it might miss some important details.

Do you think this is enough, or can we get away with fewer than this? Would 500,000 be fine? Would 50,000 be okay? How low can we go?

If we were to go as low as the number currently used for T-Rex, we would end up with about 5,000. This means that each thermometer represents about 100,000,000,000 square metres of the Earth's surface. That would put thermometers about 300 km apart. But they are not evenly distributed. In places like Africa or Siberia, each thermometer must represent a lot more space, while in parts of the U.S. each one represents less.

However, things get more complicated if you are not looking at just one time. If at each of those 5,000 locations you average the local temperatures over one year, then the range of the time-averaged values will be considerably smaller to begin with. So the statistical stability will be higher and the sampling error calculated in the same way will be smaller. We haven't really gained anything here. There are just large data hidden away with the time averages.

On this basis, an argument has emerged that perhaps we can trade off spatial detail for temporal detail. Using a small group of time-averaged readings might, so the argument goes, be like using a large group of readings at single instants.

If you wait long enough, the time average of a single site will tell you what the average over an ever wider space will be doing. If you wait for a very long time, the behaviour of the average at one station may be an approximation of the behaviour of the average over the whole planet.

In this way, the average over time would replace the average over space. There is a special name for a circumstance where averages over different occurrences of the same thing are equal to an average over time of one thing: *ergodicity*. Ergodicity is a very deep property of dynamical processes.

The time-space tradeoff involves a notion analogous to ergodicity. It is similar in that different sites function as approximate statistical copies of each other to a greater or lesser extent. So we expect that at some approximate level, time averages can help stand in for missing spatial information when brewing up some version of T-Rex.

On this basis, some people actually think we could get by with *fewer* sites to measure temperature. A number between 50 and 100 has been tossed about, if we are looking at annual time averages. What criterion is used to determine this? Why, none other than that these 50 to 100 sites (each with 365 values) can in some cases give essentially the same global mean value you get if you throw in ten or hundreds of times more sites with one instance each!

But let's not lose our way here. We are measuring a field with an infinity of values. We sampled only 5,000 values instead. And we aimed, unwisely, to replace these with a single value. All that is being discussed here is whether we can use an average across a small number of sites at one point in time or an average across a very small number of sites, each of which is further averaged over a few points in time.

Either way we arrive in the end at T-Rex once again, which isn't a physical variable anyway. For that matter, the annual time averages at the 50 to 100 sites aren't physical variables either. They do not appear in the equations of motion of the underlying dynamics. And averaged equations notoriously do not produce the same variables as statistical averages of the variables.

The time-averaged field they would be extracted from is not a "better" representation of the actual temperature field; it is just an average over fields. It is a different thing with less "noise" because the average cancels it out. But the physics of what is happening is linked to that noise, unless you have figured out how to average the physical equations too. No one knows how to do that. Moreover, averages do not fill in data that were not measured. You don't get something for nothing.

Station Closure

We mentioned earlier that for meteorological purposes, temperatures are measured by fastening thermometers in white boxes with holes in the sides. The boxes are mounted on stands attached to the ground. They are scattered all around: There is an abandoned one not far from coauthor Ross McKitrick's office, in fact. A worldwide network, called the Global Historical Climatology Network (GHCN), gathers the data from all these stations and records them for us to use later. At its greatest extent, this network had about 15,000 stations in its collection, which meant that every thermometer in T-Rex had to represent the area of a square with a bit less than 200 km on a side.

Of course there was no such network until modern times. But that did not stop the Big Panel from discussing the temperature field going back many centuries. It decorated its most recent report with a chart showing statistics dating back to ancient times, based on only 14 locations in one case and four in another. And there weren't thermometers at those places either. Yet they used these series to make confident statements about the temperature field of the entire planet a thousand years ago! This curious business will be discussed in the next chapter.

We also mentioned above that the makers of T-Rex use water temperatures for most of the world. Professor Thermos was not so far off when he put half of his thermometers in his water glass. To make the analogy truly complete, he should have put *most* of his thermometers in the glass.

For the purpose of making a temperature index, regardless of where we start sampling or what the ad hoc rules may be, it is important that we keep sampling in the same places. That is the only integrity that such an index may have. Unfortunately weather stations are moved all the time.

Worse still, many have been closed or abandoned altogether. The station near Ross's office, for instance, has a bird's nest in it instead of a thermometer. How many stations can be allowed to move or close before the continuity of the sample is compromised and the statistic must be terminated? One percent? Five percent? Fifty percent?

There are no solid rules to answer this, the ergodicity question notwithstanding. We gave up solid rules when statistics became a substitute for the physics. But critics have raised the concern that two thirds of the weather stations in the GHCN network have closed in the past three decades. Figure 4.4 shows the numbers.[5]

Why have so many closed? Stations are costly to operate. Someone has to visit each one every day to record the temperature, humidity and other data. Someone is supposed to check the equipment after every storm. The equipment is supposed to be recalibrated every few months and replaced every two years. All these things are costly. When the Soviet Union collapsed in 1990, and the Western nations experienced a recession at the same time, a lot of national meteorological agencies had to abandon many of their stations.

As you can see in Figure 4.4, the result was a dramatic change in the extent of the network. Since the "dramatic" temperatures that people worry about for global warming were recorded in the 1990s, we have cause to ask whether the change in the sample size in the 1989–91 interval played a role.

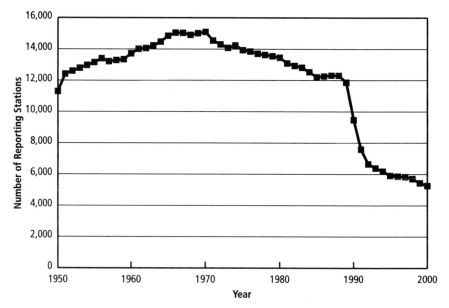

Figure 4.4. Number of stations in the Global Historical Climatology Network, 1950 to 2000.

Every time a station is closed, it introduces some kind of movement in the average. If there is a random selection of which station to close, perhaps you could hope the effect on the average wouldn't amount to much. But if the selection of stations to close is systematic in some way, there can be problems, even if you think that you can get away with 50 to 100 stations in the mean.

For instance, if originally there is a group of rural and urban stations, and the rural ones are all closed, the mean temperature will go up because the relatively cool rural temperatures are no longer used in the average. Nor will the remaining group of stations be representative of the ones closed, because city microclimates are not representative of surrounding areas.

Does this loss of measuring points actually amount to much? Perhaps not, but it is a matter of uncertainty that specialists will have to figure out. In defence of the people who produce T-Rex, they have models that are intended to take out any movements in T-Rex from changing the sample size and station locations. But that of course reminds us that Figure 4.1 is the output of a series of models nested within models, not just straight observations of something.

The only sense in which T-Rex survives is not as a statistic or an index. In reality it is not even those things. It is a sequence of different statistics grafted together with ad hoc models. When other fields employ indexes, they take care not to muddle observations, statistics and models. In fact, Statistics Canada will consciously stop reporting indexes when the rules have changed, and will call the successor statistic by another name. They did this for something called the Survey of Employment, Payrolls and Hours, from which they get wages and other labour market information. In 1983 Statistics Canada changed the way they put the survey together. They still sample as many establishments (indeed more than before), and they still measure the same things. But the rules for distributing survey forms changed slightly.

So the current data from this survey are only published in a continuous index back to 1983, and the data from the previous form of the survey are published up to 1983. The two series are deliberately not joined together. That would create the impression that it is a single continuous index. Data quality rules at Statistics Canada preclude this.

If you are measuring a real physical variable, you try to get its true value to the best of your ability, no matter what the circumstances are. But if you are calculating an index and the circumstances change, the index must be terminated and replaced by a new one. T-Rex has had it both ways. It is an index whose sampling rules changed dramatically several times, particularly in the beginning of the 1990s but it has not been terminated. Data quality rules say T-Rex must be terminated!

Sampling Altitude

One last systematic trend needs to be considered: the question of whether to use a uniform altitude for measuring temperature. Some thermometers are high up in mountains while others are near the ocean coasts. A measurement at Denver, Colorado, would correspond to an air mass one mile above New York City. Should we just throw those two together without some sort of adjustment?

Temperatures change rapidly with altitude. The temperature change between sea level and the altitude of, say, Denver is typically bigger than the most extreme estimates from global warming forecasts. There are ways to estimate how temperature would change for an air mass brought to sea level. It depends on the physical assumptions that are chosen for the trip down.

If we don't just measure relative to the surface, how should the thermometers be distributed with altitude? Should we put more thermometers where the air is denser? Some think we should distribute thermometers uniformly in space. These two schemes are not the same thing in the atmosphere and oceans. The physics is no guide because temperature is not extensive. It isn't "per" anything.

You could say that we only calculate with time-averaged anomalies, to remove the altitude effect. You subtract the mean from the temperature, right? But which mean? There are other ways to bring air down from higher altitudes along physical paths, but these do not lead to only one possible temperature at the new altitude. Would the resulting physical change in temperature from any one of these ways be the same as the change introduced by the choice of which mean you use? Probably not.

And what happens to the average altitude when a few thousand boxes get closed? You only need to move the average altitude a few tens of metres per year toward sea level to introduce as much increase in the average temperature as people say is afoot with climate change. The data are not available for us to work out how the average altitude of the network has changed. But since the network has migrated to the cities, it has likely migrated down to coastal areas as well. That's where the major cities are.

And even if those data were available, we would have no way to adjust T-Rex for altitude changes. You cannot reconstruct missing data. The birds living in the white box near Ross's office have not kept very good records.

Global What?

We have discussed some (not all) of the controversies around the methods for producing Figure 4.1. These are real debates by competent people. But to a greater or lesser extent, they miss the point: Climate is not temperature and temperature is not a single value.

As we have said, the Web site where we got the numbers in Figure 4.1 calls it the "Global Temperature." But there is no such thing as a "global" temperature. Professor Thermos explained why in Chapter 3. Physically, temperature, outside of thermodynamic equilibrium, usually in an isolating box, is not a single value. You can't have a cup of ice and a cup of coffee and report a single temperature for the two cups together. Of course you can pour them together or construct some other clever ways to get them to relax to a common equilibrium and common temperature, but different physical scripts for the relaxation will send you to different equilibria.

For the Earth, temperature is a field of widely varying values at an infinity of locations all over the Earth. Temperature is not a single value on Earth, whatever is shown in Figure 4.1. Moreover, thermodynamic equilibrium does not count time. It is timeless because nothing is happening. The very fact that Figure 4.1 reports a result per unit of time (months in this case) rules out the possibility that it measures something in equilibrium, and hence that it is a graph of something that has a single temperature.

It doesn't help to call it the global "average" temperature instead. The problem is that once temperatures are averaged, they are no longer physical quantities in the underlying differential equations. Such a thing is just a statistical construction without a physical meaning. It is an average "___" where you can make up any word you like to fill in the blank, as long as it doesn't rhyme with temperature.

On a local scale, it might be possible to get away with thinking in terms of the average, if all the temperatures are roughly the same. But what you are looking for had better be larger than the roughness of that similarity. This is not even remotely true at a hemispheric or global level. If you could travel around the world in one day, you could experience a range of temperatures spanning upwards of 80°C per day. Yet people who construct averages are telling us to look for differences of 0.1°C or 0.2°C per decade.

As Professor Thermos also explained, as soon as you add temperatures together, the number you get is no longer a temperature of anything, even if it retains units. (It retains its units in the sense that you could convert the sum from Celsius to, say, Fahrenheit consistently.) However, temperatures don't add physically, regardless of the units you use. If you put two identical cups of coffee into a pot, you have twice as much coffee in the pot as was in one of the cups, but you don't have twice as much temperature as one cup. Accordingly, adding temperatures of disjoint things, then subsequently dividing to make an average, does not recover physical meaning that was lost when taking the sum.

Think again about the cup of coffee and the cup of ice water. The coffee might be 75°C and the ice water 5°C. The mean of the values is 40°C, but that is not the temperature of either one. Now, if you poured them together into a pot and took the temperature, 40°C is a reasonable guess at what the thermometer would read, provided there was as much coffee as ice water to begin with. But even with that, the "average" could well be 20°C, if you are prepared to wait until the mixture relaxed the whole way to room temperature. It all depends.

This kind of average does not tell us anything about how to interpret temperatures in the world on a day when Antarctica is −40°C and India is +40°C. No one is proposing to pick up the continent of Antarctica and move it to India to mix them together. The mean of temperatures in those regions describes neither place. The mean temperature between the cup of coffee and the cup of ice water describes the temperature of a liquid that, at the moment we took those temperatures, does not exist and may never exist. Changes in that mean over time necessarily describe the change over time of a non-existent thing too.

The Simulacrum Fallacy

Let us dispense with a misconception. If we construct a statistic and claim that it represents something, the claim and the statistic do not cause that thing to exist in the world if it did not exist previously.

Just because you can construct a number to represent something, and even attach a label to the number that corresponds to what you are trying to measure, that does not make the number meaningful. It does not conjure into existence the thing you are attempting to measure. If the thing did not physically exist in the first place, you can't make it exist by assembling enough data. To treat a conveniently defined statistic as proof of reality is an example of what we will call the *simulacrum fallacy*.

The word "simulacrum" ordinarily means a representation of something, but in philosophy the term is now used to mean a copy of something that does not exist, or an image of a non-existent referent.

A slight variation: If at first it doesn't exist, get more and more data and say it does anyway. However, this amounts to the same thing in the end. More data won't help.

To compute a mean, you first add up the things on the list. If the process of taking the sum disconnects you from the concept behind the numbers, the average's ultimate meaning is in doubt. Professor Thermos made a joke about computing an average telephone number for his colleagues to show that an average of worthwhile numbers can have no significance.

From a global perspective, this example would lead to a global mean telephone number. Yes, there is a global mean telephone number! And, no, to our knowledge, no one has computed it. What would it mean if someone did compute it? It is not in any sense the most "representative" phone number, nor is it the number that all phone companies will tend toward assigning. Nor is the person whose phone would ring if you dialed the closest actual number to it the most "typical" phone user.

We can also compute the global median telephone number. We could compute the global root mean square telephone number, just as Professor Thermos did for the temperatures. But none of these averages, nor any of an infinite number of others, would function as a measure of central tendency the way "average rainfall" or "average income" does.

All of them just pick out one number to replace a collection of numbers. Sometimes the collection is all there is and no one number can meaningfully represent the collection any better than any other. That is what happens in the physics of intensive thermodynamic variables like temperature. You may have a range of values, but nature doesn't tell us which one value stands for all the rest.

No phone number represents all of the others. The sum principle is just an example of the idea. The sum of all telephone numbers has no meaning, even though you can compute it. A final division to complete the mean doesn't change that.

Maybe the global mean telephone number can be used to set a meaning for global temperature. Call the mean telephone number and see what the person who answers thinks the temperature is. Why not? The PUN report approvingly describes a survey of people to forecast future climate and another survey to determine the uncertainty of the forecasts.

Calling T-Rex the Global Temperature is an example of a simulacrum fallacy, because regardless of how you put the numbers together, there is no such thing as a "global" temperature. It all depends on what the averaging rules are. The mean, which is widely called the average, is only one possibility for rules to make an average. Without some physics to tell us which rules to select, any temperature in the field can represent the field as an average. This, by the way, can be shown mathematically.

While every possible average exists mathematically, no average exists with specific meaning in terms of the thermodynamics. No one "average temperature" appears, as far as anyone knows, in the governing laws of physics that apply to weather and climate. So calling Figure 4.1 the Global Average Temperature is a simulacrum fallacy: estimating a physical value that does not exist.

What Does T-Rex Have to Do with Climate?

All these troubles—science, statistics, and even existence—yet people talk about T-Rex as if it were as routine as reading their weight on the bathroom scales. They think it is a temperature, but T-Rex is really an arithmetical construction that puts the bite into the term "global warming." If T-Rex does not actually exist physically, does global warming exist? Should we be more sophisticated and talk only of *climate change* to do justice to the subtleties of the science?

Notwithstanding all this, journalists report on every move of T-Rex, panels of scientists study it, teams of experts do mathematical analyses to track changes down to the hundredths of a point, and international treaties commit nations and their governments to do something about it.

The conventional rule is that if T-Rex goes up one Bleep, that's okay. If it goes up two Bleeps, that's not okay, and if goes up three, then you're out and the game is over. Why? No one is quite sure; that's just the rule.

Think of the coffee cup and the ice water again. Initially they are at 75°C and 5°C. Room temperature is 20°C. What sorts of changes do we expect over the next few minutes? The temperature changes in those liquids will be driven by gradients, or differences between the state of the liquid and the state the physics is telling it to go toward. Other things being equal, we expect the temperature of each liquid to move to 20°C, so the mean will go down. In other words, the two cups will in some combined sense go through an average "cooling."

Suppose we come back in 10 minutes and find that the coffee is now 72°C and the water is 10°C. The world is unfolding as it should: The coffee is cooling and the water is warming. Yet despite the fact that we expected the average temperature to go down, the average temperature is now 41°C, so between the two cups we have experienced an average "warming" of +1°C.

This is entirely possible. The temperatures say little or nothing about the process by which they relax to a single value. The coffee is cooling a bit more slowly than the ice water is warming. Maybe the water glass is sitting in a patch of sunlight, or the coffee cup is insulated. We didn't say.

Or maybe we ought to have computed the averages differently. We could introduce an infinite number of bona fide averages that would have gone down, so that expectations of ultimate cooling could be satisfied. This is a mathematical fact. There could be any number of reasons for the observed change, but it is not a matter of concern. The thermodynamics only tells you where a thing's state is going to be eventually. It does not tell you when the thing will be at that state or how it will get there. It's like a very bad rail service.

What does it mean, then, if the average "anomaly" as shown in Figure 4.1 goes up or down over a particular stretch of time? It really isn't clear at all. The underlying physical dynamics that determine what the next values in Figure 4.1 will be don't take the any values of T-Rex as an input. What T-Rex does next is not contained in T-Rex in any way.

The connection between T-Rex and the Earth's climate has always been contrived. When the first radiative-convective climate computer models emerged before GCMs, in the early 1960s, the Earth was represented in such a simple way that it had a single temperature at its surface. The model didn't have any other temperatures at the surface, so for that model there really was a "global" surface temperature. That is not how the real Earth is put together; it was just how the model was put together.

There was of course no way to relate that temperature to T-Rex in any theoretical way, although they were uncritically equated with each other in the minds of many. The earliest of these models made some people think that T-Rex would increase by about 1 B if the amount of carbon dioxide were doubled. This was not regarded as significant. Don't ask us why. People didn't ask such questions. When increasing carbon dioxide was also made to increase water vapour in the revised model, the effect doubled, leading to a 2 B increase in T-Rex.

Leaving aside the matter of how water was used to amplify the effect of carbon dioxide doubling, the second increase was regarded as significant. Why was a 1 B change viewed as unimportant, while one of 2 B was regarded as important? Why would such a small difference between such model warming experiments make any difference in terms of global climate change?

No one attempted to answer these questions. It just became a convention: 1, good; 2, bad. There was good reason not to ask why, of course, because no one knew how to physically connect this value to what was actually happening in the real world, place by place in the oceans and atmosphere. As we have said, temperature, outside of an isolated container (assuming temperature exists meaningfully at all), is tied to locations, and it is not global.

If you have an ice cube, it doesn't matter that the average winter temperature outdoors may be –5°C (23°F) if the ice cube is indoors by the fire. The thermodynamics of its fate is tied to the temperature of what it is exposed to, and *not* to some average over temperatures from elsewhere and times past.

Having decided that a rise of 2 B is bad, governments are now spending a lot of money asking scientists to figure out what is bad about it. As no one knew what connection 2 B had to global badness, it was necessary to start thinking up connections.

Many went to work on this. Efforts have gone on for years in the form of "impacts" research. This research has had at least one exciting effect. Before it, impacts were known as "consequences" or "effects," while "impacts" referred to the force of a collision (materials under impulsive loads, to be technical). Now, with the new terminology, climate change research has become as action-packed as a Superman comic book: Bang! Kapow!

An example was a recent study arguing that global warming would be bad for children's health in Ontario. The reasoning was that global warming means T-Rex goes up, and if T-Rex rises, then it is hotter everywhere all the time (bang!), and if children get overheated, they have to go to the hospital and that's bad—kapow! The folks who run the government of Canada's Climate Change Action Fund were really impressed with this study and displayed it on their Web site for a long time. We will tell you more about this action-packed work in Chapter 8.

Many contemporary prognoses in terms of T-Rex call for a change double again that of the early radiative-convective models, or even a bit more. But why would that be any more compelling than 2 B was? The change they describe in Bleeps remains very small compared to the variability we have experienced in degrees C on the Earth, and it is minute compared to the number that physical formulas use in terms of absolute temperatures. Why are those numbers any more significant for climate change than 2 B was? These questions are not answered scientifically.

Nonetheless, some might say it would be hard to rule out significant changes if there is a change in T-Rex of +4 to +6 B. It's true. But it is misleading to say so and stop. It would be hard to rule out significant climate change even if T-Rex did not Bleep up or down at all. We could have a zero-Bleep climate change caused by humans too, and it could be severe and catastrophic and everything. Burning fossil fuels could even cause it. All this without any change in T-Rex!

How could that be? Easy. We can have the local weather move toward different temperatures with changes that cancel each other out in the sum of the temperatures that people make before they divide by the number of temperatures measured. You don't have to go toward more extreme values even, just scramble the locations of current local climates to create substantial catastrophes. You don't even need to scramble things; even a simple systematic shift of weather by a few degrees of longitude should create a big mess. Unfortunately, because of the deficiencies in the global observation net, such a shift might still show up as a change in the official average, even though it would be exactly the same temperature field just shifted in longitude.

You could have both scrambling and some systematic movement to different local extremes. We could have more clouds in some places and less clouds in others—at the same time! We could have more winds and rain, and more drought and more snow, and less of everything too. We could have more glaciers or less glaciers—maybe a tasteful mix of both. We could have the sea levels rise, or fall. All this could be done even as T-Rex coasts along unchanged. Climate change could just find a different way to express itself.

All of the portents, omens and signs that people point to as consequences if T-Rex goes up can also happen if it doesn't change at all. We can even have more hot days in the summer or warmer winters and still have no change in T-Rex. The deficits can be made up where we don't have thermometers, which is nearly everywhere. Or you can have local changes in the daytime maximum and the nighttime minimum that just offset each other in the simple mean.

All the great old portents, omens and signs can happen even if T-Rex goes down. Sea level can rise even if certain global average temperature statistics go down. These portents are sensitive to local physical circumstances that can have very little to do with a simple sum of some temperatures from elsewhere and elsewhen.

The Ice-Cold Hard Facts

We used the example of an ice cube indoors by the fire in the winter to distinguish averages from relevant physics. This is not really a contrived example when you consider glaciers.

Glaciers are the big ice cubes outdoors. They can shrink even if certain average temperatures drop. They are dynamical things because they are not in a closed box but participate fully in their environment. Their size depends on the properties and timing of many physical and thermodynamic conditions (like cloud cover and precipitation, to name just two) that give them surprisingly complicated behaviours. Pointing a finger at a glacier and saying it is getting shorter so therefore the temperature is going up is naive and very convenient for the Doctrine.

There is a long tradition to finger-pointing at glaciers. In a recent issue of the *New Zealand Herald* (April 17, 2002), there was a news item warning about melting glaciers in the Himalayas. It began:

> The giant glaciers of the Himalayas are melting so quickly that
> within five years dozens of glacial lakes may burst their banks,
> killing tens of thousands of people in their path, a United
> Nations report warns. Researchers from the UN's Environment
> Programme, which has been monitoring the glaciers for the
> past 15 years, are deeply alarmed by the speed with which
> many lakes are filling up as glaciers shrink because of global
> warming.

Notice that last bit: *because of global warming.* Other news coverage around the world echoed the same point. Later in the same article, the director of the UN Environment Program weighed in, arguing that the threat of deluging villages downstream from these glaciers makes it imperative to begin reducing global carbon dioxide emissions. (Fortunately, engineers in Nepal are being more practical by reducing upstream lake levels.) So there is the claim: T-Rex is causing melting glaciers.

Surely temperatures from places like Miami and Berlin, which contribute to T-Rex, have nothing to do with glaciers in the Himalayas. Remember the ice cube by the fire. What does the glaciers' situation have to do with a global average of temperatures?

They sit right in the middle of the lower troposphere. You can look up the lower-troposphere temperature records at the Global Hydrology and Climate Center Web site, www.ghcc.msfc.nasa.gov/temperature/. On the map of the world, use your cursor to draw a box around the Himalayas, roughly 26.25°N to 33.75°N and 86.25°E to 91.25°E. The Java program will show you an average temperature of the air in that region since 1979. The average "anomaly" in 1980 was +0.18 B and in 2001 it was –0.07 B, and between these years there was no increasing trend whatsoever. Oops, no global warming going on there in the average. Perhaps the temperature discreetly blips up, escaping the average in between measurements, to get in some quick melting when the instruments aren't looking.

We would hate to see Nepalese villagers and others in the path of glacial lakes come to harm as a result of sudden flooding. But if it happens, it is not because of rising average atmospheric temperatures, because this average is not rising, and has not been for at least the past two decades.

You can see why we become skeptical of big UN panels and their alarm-ringing pronouncements. They take an observation of increasing lake levels in high-elevation sites in Nepal, hop to the conclusion that the glaciers are melting due to rising average temperatures, skip to the conclusion that "greenhouse gases" are causing it all and jump to the announcement that we need to cut fossil fuel use. Yet they did not bother to find out that the temperature field around the glaciers does not show any sign of increase, on average. Glaciers are complicated things. They are not good thermometers.

One Strange Beast

We cannot call Figure 4.1 a "global temperature," because there is no global temperature. Nor can we call it the "global average temperature," because once you average temperatures, they are no longer temperatures; they are only statistics.

Can we call it a temperature index? No, it even fails the test of being an index! The rules for putting together an index couldn't be easier: Just toss whatever you want into the mix and average them however you like. The only stipulation is that whatever rule you applied last year, you need to apply this year, and next year, and every year after that too.

But as far as temperatures go, the rules applied in 1901 are nothing like those applied in 2001. The temperatures we gather today are collected in different ways in different places, using different methods than even 10 years ago. The weights in the temperature average change regularly, reflecting the different densities of thermometers in each region over time. Even if the continuity of the index could be said to have held up from 1900 to, say, the 1970s, it broke down at 1990. Indeed, there are many places in the series where the continuity is compromised. Hence Figure 4.1 is not a temperature index, it is at best a sequence of different temperature indexes, all spliced together, that should have been repeatedly terminated.

This is not really a problem in the end. No one really cares about T-Rex. We don't live in an average temperature any more than the ice cube melts in one. What people worry about in connection with climate isn't averages over temperature. No one cares if the temperature were to go up a few degrees in the mean.

They worry about floods, droughts, storms and other catastrophes. But this is not new. The risk of these has always been a worry for people, even long before "global warming" came into our lives. No one has made any meaningful connection between the non-physical average that T-Rex is and these things. There is much fear that there is a connection, but there is no science that we have been able to find that stands up to scrutiny.

In that light, you can call T-Rex what you want and make of it what you wish. But as far as we can tell, this T-Rex is one strange beast that has little or nothing to do with the climate that any of us actually experiences.

---

[1]   The Hadley Centre, UK: www.met-office.gov.uk/research/ hadleycentre/obsdata/globaltemperature.html.

[2]   Christy, John R., David E. Parker, Simon J. Brown, Ian Macadam, Martin Stendel and William B. Norris. "Differential Trends in Tropical Sea Surface and Atmospheric Temperatures since 1979." *Geophysical Research Letters* 28(1), 2001, 183–86.

[3]   The satellite method is introduced in Spencer, Roy, and John R. Christy. "Precise Monitoring of Global Temperature Trends from

Satellites." *Science* 247, 1990, 1558–62. Updated data are available at http://vortex.nsstc.uah.edu/data/msu/t2lt/t2ltglhmam.d.

4   See Ebbesmeyer, C.C., D.R. Cayan, D.R. McClain, F.H. Nichols, D.H. Peterson and K.T. Redmond. "1976 Step in Pacific Climate: Forty Environmental Changes Between 1968–1975 and 1977–1984." In J.L. Betancourt and V.L. Tharp, editors. *Proceedings of the 7th Annual Pacific Climate (PACLIM) Workshop, April 1990.* California Department of Water Resources: Interagency Ecological Study Program Technical Report 26, 1991, 115–26.

5   We are grateful to Joe D'Aleo, a meteorologist at Intellicast Corp., who retrieved the data from the GHCN, analyzed it and shared his results with us.

5 T-Rex Plays Hockey

The Mutant Ninja Temperature

In the last chapter, we joked about T-Rex being like a cheesy horror movie monster conjured up in a lab, who terrorizes the town until the superhero arrives to stuff it in a cage. Movie monsters are always more fearful if they can be described as "mutants." It makes them seem more hideous if some strange incident, usually involving radiation, has transformed them into something evil. Regarding climate change, we have the strange incident involving radiation: We burned fossils fuels and added some infrared-absorbing carbon dioxide to the atmosphere. All we need now is to show that T-Rex has mutated and turned into a powerful menace.

Proponents of the Doctrine have for decades been looking for a "smoking-gun" proof of this mutation. The whole idea is a slippery one from the standpoint of physical temperature, as should be abundantly clear at this stage of the book. Chris was asked once to consider devising a smoking-gun proof that would involve measuring infrared radiation at the surface. Unlike approaches focused on T-Rex, at least that was based on something with physical merit, even though the idea never worked out.

However, there is something that has been put forward as a smoking gun within the last few years that looks pretty convincing if you do not know about the origin of T-Rex and the nature of models. It is a set of temperature numbers dating back to the year 1000, which we have plotted in figure 5.1. It shows averaged temperature "anomalies" and the units are called "degrees C". The shape has wonderful shock value. It is often described as the hockey stick because it is long and flat like the handle of a hockey stick and it bends up sharply at the end like its blade.

You can see the mutant temperatures starting around 1900. This

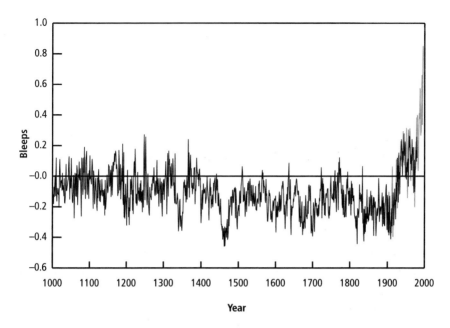

Figure 5.1. The hockey stick graph.[1]

image has become the poster child of the Doctrine. It appears, in colour, in nearly every venue where people argue on the Doctrine's behalf. In Big Panel documents, it occupies almost half a page in the SPAM, more than half a page in a separate report called the Technical Summary, and a full page in the main text of the Big Panel report (actually it appears twice on the same page). In yet another summary called the Synthesis Report, it appears not once but three times. There are probably card-sized versions now in bubble gum packs. It is the Official Statistic for the past 1,000 years.

On the basis of the hockey stick graph, many people refer to 1998 as "very likely" the warmest year of the millennium. Big Panel scientists believe it shows 1998 to be the warmest year for the whole world since AD1000, and they have said this before the U.S. Senate and in other government briefings, on national radio and in national newspapers in Canada, in science journals like the *Proceedings of the National Academy of Sciences* and everywhere the PUN makes one of its high-level briefings. Environment Canada recently published full-colour pamphlets in Canadian newspapers repeating this claim.

The hockey-stick diagram has taken on a remarkable life of its own, which is probably something the authors of the study itself never intended. For one thing, they never claimed it shows the "global" temperature, only the average of the northern hemisphere. That claim is of course problematic in its own way, as we will discuss. But we have to comment that while in the reports they point out that it refers only to the northern hemisphere, in many public settings PUN leaders refer to it as a global statistic. In other words, they wave away the distinction between the whole world and half the world. Once people start letting details like that slide, there is no telling the mischief that can follow.

This whole chapter will be about the hockey stick. We focus on it at this length not because we think it is important, but because the Big Panel does. Our purpose here is just to show that this kind of study, regardless of how well or poorly it is done, cannot serve the purpose it has been asked to serve, namely, as incontrovertible proof that we face a global warming problem.

To begin with, it is a diagram that refers to the past. But the world is troubled by what might happen to the climate in the future, not what happened or did not happen over a thousand years of pre-industrial history. The PUN could be completely correct in warning about a future crisis even if the hockey stick were as flat as a prairie road.

The hockey stick only reveals something about the future if you assume that behind it is heaped the ruins of some scientific theory asserting that global warming is not a problem, to which the hockey stick is a shattering rebuttal. People typically assume this whole structure of argument has been made, somewhere, by someone, and that when the Big Panel presents the hockey stick, they are just showing the punch line while keeping the rest of the argument on file back at the office. Unfortunately, there is no argument on file. It is just a graph, and if you are going to conclude something from it, you need to understand for yourself what it does and does not teach us.

As a prop in the background of presentations or articles, the hockey stick is very effective. It is rarely subjected to any critical reflection, and even the people who use it to make a point rarely give any thought to the question of the difference between a temperature statistic and temperatures. They never wonder whether ships from the late Viking age really were equipped with thermometers. It does not occur to them that in the

Medieval period people did not trudge miles away from their village into the surrounding woods, locate the spot where an international airport would stand seven centuries later and record the daily temperature. In fact, no one was taking temperatures anywhere until Galileo and his contemporaries invented versions of thermometers near the end of the 16th century. And it wasn't for a couple of centuries more that any but the sparsest temperature records, never mind "global" ones, began to be kept. Yet somehow we have an unblemished and unbroken temperature statistic back to before the Norman invasion.

The first question people should ask is, "How did they do that?" But the sharp turn of the blade of the hockey stick at around 1900 is so abrupt and startling, not a thought about the technique is whispered. The Doctrine's pitch plows under any such questions. No one seems to marvel at the hyperbole generated by this graph, only at the graph itself. In this plot, the Doctrine has found its smoking gun.

We will show you that it is not a smoking gun at all. Nor is it evidence that there is anything remarkable about our era. The exasperated reader at this point will ask how we can say this, with temperatures sharply changing about the time the industrial age was seriously getting under way? The answer is very simple: The plot is an artistic pastiche of two different things. The blade of the hockey stick is one thing and the handle is something entirely different that has been grafted onto it. It is the artful joining of these entirely unlike things that makes it possible to talk about temperatures before thermometers were invented, and the joining is precisely at the point where the blade of the hockey stick turns upward.

This joining certainly wasn't done to produce a deception. It is the result of scientists poking around with accumulated data. They were trying to make sense out of it, and if that meant putting things together in this way, we say more power to them! However, if this type of enterprise is taken out of context as some kind of definitive result; if it is taken as producing actual temperatures measured in degrees instead of the concatenation of two unlike statistics pushed unnaturally into having temperature units together, then the Doctrine is strengthened at the expense of real understanding. The graph actually represents, as in most science, a struggle against our own limitations and ignorance. In this chapter, we hope to free it up to serve that purpose by helping you understand its real strengths and deficiencies.

The Rings of Power

The handle of the hockey stick does not originate from thermometers at all. It is based on data that are called proxies. These proxies are supposed, optimistically, to be the next best thing to having thermometers without actually having them. They include, among other sources, the thickness of tree rings from very old trees, data from coral growth, old pollen counts and analyses of oxygen in the layers of very old snow within glaciers.

The proxy data are fascinating records, but what they tell us is a complex and subtle story we may never fully understand. We need this sort of data, but we must fully accept essential limitations in using it.

For each type, there is a long chain of reasoning between the actual measurement and the one-dimensional character called local temperature with which we are so obsessed. For example, in a tree ring, there is a question of whether to measure the ring width or the wood density within the ring, since each can be affected by surrounding temperatures but in different ways. And the tree yields only a single measure per year (actually per summer, when the growth occurs), but many things can happen within a year. Two years that look similar, ring by ring, can be different in many ways. Conversely, rings that look different can be generated in years with the identical temperature statistic values. This is especially true over the long time spans represented in Figure 5.1. Trees grow in concentric rings, and if the tree were growing in a perfectly constant climate, the rings would nevertheless get steadily thinner and thinner just because of the geometry of a tree. That thinning must be accounted for and mathematically removed before using the ring thickness as a temperature proxy. This process is called de-trending, but a tree does not come with a set of instructions about how much de-trending is needed, or whether the amount of de-trending needed is linear or nonlinear or very nonlinear. There is judgement involved, and more than a bit of guesswork.

If the ring were to represent temperature alone, it would be a local temperature "biologically" averaged over a year. But biology is more complicated than temperature too; otherwise, we would only need heaters to make ourselves good gardeners. All that bother over water, soil, sunlight, wind, insects and more could be dispensed with. That bother over the other things is what climate is all about. Temperature is just one member of a large chorus.

So scientists try to select among the old trees those that give some reason to believe that temperature is the major influence. There are no guarantees, but they do the best they can. Selecting only some from a very small range of choices makes for very few candidates. The few trees that are suitable tend to be in harsh and remote places far removed from one other.

Using the data from these trees to generate the history of temperatures where we currently measure temperatures depends on our ability to relate these trees and their rings to modern local-temperature records. One of the questions people ought to ask when studying a picture like Figure 5.1 is what do tree rings, or even temperatures in the air near those tree rings, in remote locations dating back to before the Middle Ages, have to do with temperatures at, say, O'Hare Airport in Chicago in the late 20th century? How can such airport temperatures from all over the world represent a continuation of temperatures felt thousands of kilometres away, centuries before, in remote woods?

It's an amazing scenario. It's like saying the weather in Calgary at Aunt Mabel's June wedding was a continuation of the temperatures at Khartoum in November 1612. The scenario doesn't become less amazing when you add in more types of proxies.

The other proxies each have basic problems of their own. And all of them are supposedly linked, like apples to oranges, to unlike physical things, supposedly linked to each other and to late-20th-century airport temperatures. They all pertain to local temperature, if they connect to any temperature at all. But just because they are proxies and not temperature measurements doesn't mean they are spared the problems of relating local properties to global statistics. Moreover, the global coverage of the proxies is extremely sparse compared to the modern temperature networks. Averaging their effects constitutes a very different global statistic than the mean over our current temperature-measuring network. The difference is not only a difference in kind but one in coverage in both space and time. There is little physical reason to believe it would behave the same as T-Rex.

Maps and Mappings

Then why do they agree so well where they overlap? It is not because they just naturally do so. They were made to agree!

Statistical methods are used to graft the proxies onto the temperature data, like a gardener attaching the limb of one type of tree onto another.

The proxy data are subjected to a mathematical process that produces new numbers, with units optimistically called temperature. But these new numbers are not measurements of temperature or anything else. They don't have units of temperature. There is just some mathematical rule cooked up using those methods as scientists poke around with the data that take in proxies in a wide variety of units and spit out numbers labelled "temperature."

There is nothing wrong with scientists trying out some rule like that. Ideally, there should be a good reason to believe that there is a physical connection between the things being joined. The danger is that there may be no connection of any reasonable kind between them. The statistical methods won't warn you about that. They don't warn you when things are going wrong. They always work, even when you would rather they didn't!

It is not just statistical methods that can have this dangerous property. Any type of data can be matched with any other type of data, exactly, given some experience with mathematics. If you have some observational data and you want to make a rule to take every datum to a datum in another data collection, you can find one. You can find lots of them. There is an infinite number of such rules. Every rule will work perfectly for the data you have set it up for, but they will all behave differently from each other when applied to new data that were not part of the data used to make up the rule. The only rule that will work for all data will be one from a basic physical theory. Making up rules to match data is not basic theory. It is called *fitting*.

Any one rule between values is called a map in mathematics. A map tells you where you must go, depending on where you are. If you are at a particular datum, a map will tell you what datum to go to in the other collection of data. You start at a point and are directed by the map to a corresponding value in the other data set. That is what the map does. It is the rule that you use to go between two collections of points. There is no scientific basis necessary. It is just mathematics that you make up for the occasion.

This is less complicated than it sounds. You use mappings all the time. There are road maps, of course, but other kinds of mappings are used whenever you connect unlike things. *If it's Tuesday, take out the garbage.* There is no law of physics that says, "Day equals Tuesday" implies "garbage at curb," it is just a mapping rule. You could take it out on

Monday and let it sit for an extra day. A mapping rule might say, "Tree ring width on Mount Washington is 0.128 cm" implies "average temperature in northern hemisphere was 14°C." This is not a law of physics or even biology, it is just a mapping that fits together two pieces of data the researcher wants to stick together.

Mappings in science are used all the time on data, but there is a danger. When the rule is developed, there has to be some check on the results. If the mapping starts producing numbers that we know on other grounds cannot be correct, then the rule is breaking down. Its proper limits have been breached, and that is something worth knowing.

Things are more dangerous with maps produced from statistical fitting. Statistical fitting means that the map has two parts: the fitted part and the "residual." A residual is just a leftover bit showing how far the mapping missed the target. It is chalked up to random deviations. Using statistics means the fit is more relaxed and less brittle.

Brittleness can be good, as the shattering of the thing can be a warning that the method has outrun its usefulness. In a statistical mapping, there is nothing to tell you that the fit fails in any particular case, because any deviation from an accurate mapping can be attributed to randomness. If you say that there are other physical properties to be considered than what the map is designed for, and that the two things being joined don't really connect strictly to each other, you can cheerfully say that the random deviations take account of all of the other properties.

That is the case with the mapping between temperature and tree rings. You would not expect that temperature alone explains the tree rings. But with statistical fitting you can say it does anyway, and chalk up all the rest—sun, water, soil characteristics, etc.—to the random deviations. It's not pretty, but it's done.

If you look at the plot of the hockey stick curve just where the handle meets the blade (Figure 5.2), you can clearly see random deviations. The hemispheric temperature statistic is what the proxies are supposed to be fit to, but observe that the map does not actually produce the same value as what the map is aiming at. Not one of the fitted values agrees with the target values. If a non-statistical fit was cooked up, the numbers would agree exactly. There would be no difference at all.

This can be seen in Figure 5.3, which plots the differences themselves. The differences are not small compared to the changes in the hemispheric

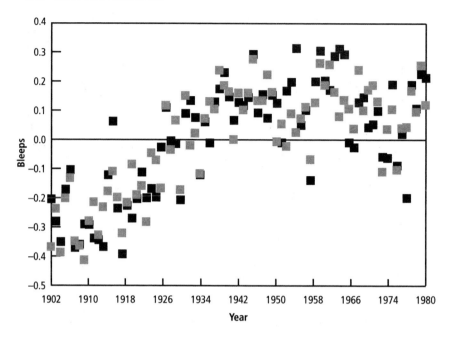

Figure 5.2. Data points during overlap interval. Black squares show northern hemisphere T-Rex (the target variable) and the gray squares show the numbers produced by the mapping.

temperature statistic. It tells you that the map can only reproduce temperatures to within ±0.3 degrees. This is not very precise, considering the temperature changes we are looking for. But there is lots of leeway to explain away any inconvenient deviations or troublesome and undesired physical dependencies.

This degree of deviation also allows the map to be constructed as a linear rule. With a bit of randomness to act as a sort of mathematical lubricant, you can fit linear maps to nearly anything.

If you look at the actual publication behind the hockey stick, the map that is used involves something called a singular value decomposition, which is a matrix factorization from linear algebra. It is well known as an effective method for doing things like data compression, which is useful in computer applications. It is a good tool for that job, but things are more problematic here.

The particular statistical methods used here cannot tell us if the mapping is physically reliable or not. But statistical mappings can be evaluated

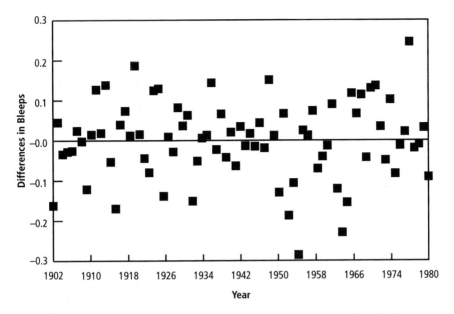

Figure 5.3. Differences between target values and proxies after fitting series together.

by comparing a particular average of the fitted values to a particular average of the target values. The formula for these averages takes the sum of the squared values: This is another one of those cases where theory does not imply a linear average, or simple mean. The formula yields a ratio that, in the jargon of the trade, tells us how much the target values are "explained" by the mapping. Of course it is not really an explanation of why the target values are what they are, since it is not a theory. The term is just a bit of jargon.

To perform this evaluation, the authors did the fitting on only one segment of the temperature statistic data, spanning 1902 to 1980. But they also had temperature statistic data for the period from 1854 to 1901. So they used the mapping rule and the proxies to "predict" the temperature statistic back to 1854. Then they could test how much of this earlier period is "explained" by the mapping and the proxies.

Altogether the authors had 112 proxy series, but most of them only go back to the 17th century. Only 55 of them extend back before AD 1600 and only 12 extend prior to 1400 (of which only eight are actually from the northern hemisphere). Since the mapping takes the proxy data as an input to produce a T-Rex output, a different mapping rule is needed for each time segment, with different numbers of proxies and different amounts of randomness in each.

In the original paper in which the hockey stick method was explained,[2] the authors report that the mapping derived using all the proxies "explains" only about 30% of the temperature data in their collection. This set only extends back in time to 1820. Prior to that, the set of proxies shrinks and the proportion of the temperatures they can explain also shrinks. Quite rapidly, as it happens: The pre-1820 proxies "explain" only 5% of the temperature statistic, and the 12 proxies that run back prior to 1400 "explain" zero percent of the temperature statistic in the pre-1901 interval. That is why, in the original paper, the authors truncated the temperature statistic constructed by their mapping at AD 1400. In another paper a year later, they did some analyses on the pre-1400 portion and consequently drew the whole sequence for the reader's benefit. That is the picture the Big Panel put into the SPAM. They included a 400-year span for which the mapping is known to be no better than random numbers.

Wishing for Stationarity

The authors of the hockey stick papers used a somewhat involved technique to generate their mapping. But there is no unique way to construct a linear map of this type. You don't need to use fancy linear algebra here. Specialists argue about these things all the time, but that is not important to us. What is important is that all maps, linear or otherwise, have within them constants that are discovered in the fitting process. These constants are not fundamental constants of nature. They are not the speed of light or the rest mass of an electron, for example.

If we use this map before or after the times in Figure 5.3, we are no longer just mapping one data set onto another mathematically; we are using the map as a model. This brings us back to the issues in Chapter 3 about distinguishing theories from models and controlled experiments from observations. The temperatures from before 1900 in Figure 5.1 are a result of a specific linear statistical model applied to observations. The entire handle of the hockey stick is the output of a model *extrapolated* into the past. Those temperatures are not based on a physical theory arising from controlled, repeatable experiments. A different model would give different results.

So not only within the model itself do you not know the temperatures to within about ±0.3 degrees, you cannot say where the model will go wrong before 1900. The entire hockey stick handle is an extrapolation that could take any number of shapes other than what it is depicted. The deviations between actual temperature means and the values seen in Figure 5.1 could be much more than ±0.3. There is no way to determine how large the error could be. If these deviations could be plotted over the 1,000-year interval, as long as they are vanishingly small in the interval after 1900, nothing in the fitting procedure tells us how big they are everywhere else.

To understand what this means for the map, imagine that you travel back in time to all the (mostly wilderness) places where weather stations are today to fill in the missing data from the past. With these data, using exactly the same method, we could compute new mapping coefficients for different time intervals. We could compute the map using data from 1351 to 1430, for instance. Or 1711 to 1790. The constants for each one of these redos would undoubtedly be different from those of the map used to make Figure 5.1.

Why should they be the same? There is a curious tradition within fields using this sort of methodology to reverse this question and ask, why should they change? There is a practical reason for that traditional reversal: Life is simpler that way. But it is not truer. If the constants for all the intervals were the same, the sequence of values would be called stationary. If it is stationary, you do not need to step out of the statistical model and ask the difficult theoretical questions that non-stationarity would imply. **Stationarity** is not a bad assumption in the absence of anything else. Neither is it a good one, but in any case it is an assumption—a big one.

The only constants that we do not expect to change are the fundamental physical constants, as mentioned. All the rest, including the coefficients of mappings, will change. Only how and when are in question. That is one of the reasons that extrapolation is generally suspect. The constants are likely to be functions of the climate details. Furthermore, they will be properties of the proxies selected, how they were analyzed and the network for measuring temperatures. They will also be properties of the rule used to compute 20th-century temperatures where there are no thermometers. Change these things and the constants will change, even if the climate itself does not.

Even if all other things are held fixed, the network for measuring temperatures has changed. If you were to travel back in time and place thermometers to record the missing temperatures from the current network, you would find the network shows local climate change just because cities have grown up around the measurement sites. Cities are known to have higher temperature than the surrounding countryside. The effect is known as the urban heat-island effect. While this is an interesting issue in its own right, it practically guarantees that local climate change at the points of the temperature network has taken place, implying non-stationarity in the constants of the map. This effect would be absent from the proxies, so it would require a map that adjusts for this artificial drift of the temperature network with respect to the proxies.

There are deeper reasons for a drift in the map. The constants would change slowly, drifting to new values, which reflect different statistical relationships between local temperatures and the non-physical hemi-

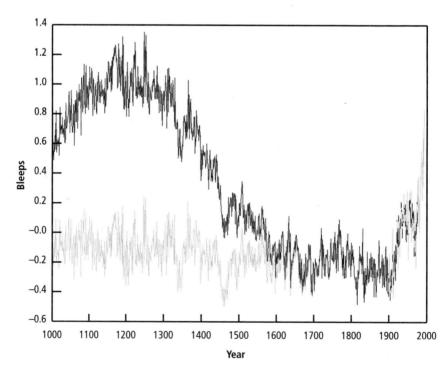

Figure 5.4. The hockey stick curve with a simple parameter drift process added.

spheric average temperature statistic. This would cause the same proxy data to correspond to different hemispheric temperature statistic values. In Figure 5.4, the top line represents a mapping with a climate drift built in. The adjustment is such that the presence of the drift cannot be detected prior to 1600. Only in the Middle Ages does it show itself.

How did we get this adjustment? We simply made it up. We added a ratio of polynomials in time with the correct asymptotic behaviours. For those who find that mathematical-sounding, let's just say it was a garden-variety adjustment of no particular extremity. Nor is it anything that the map-fitting process could rule in or out.

But some might object that this is an extreme deviation compared to random changes in the hemispheric temperature statistic. They would be right—sort of. It would mess up this smoking-gun claim of the Doctrine that we are experiencing the warmest temperatures in a thousand years. But for those who may think our invention is in any way extreme in a physical sense, they should think again.

Figure 5.5 is a replotting of Figure 5.4 with the temperature statistic presented as if it was a real temperature that the formulas of physical theory will take. These formulas all take absolute temperatures in degrees Kelvin, not Fahrenheit or Celsius. The ideal gas law, the Stefan-Boltzmann law, or Navier-Stokes all take temperatures in this way. Phase transitions are a potential complication, but the values we are speaking of are far from those. They do not come into play for this statistic made physical. Figure 5.5 shows the various temperature curves the way nature sees them, not in the arbitrary "anomaly" form in which the Big Panel likes to show them.

In the research field, where proxies are used to construct temperatures, absolute temperature scales are rarely put up the left axis. The plots are invariably done in terms of "anomalies," a term we explained in the previous chapter. In Figure 5.5, we show it in terms of absolute temperature and not anomalies. The issue of the Doctrine pertains to physics and not to the statistical behaviour of the series of temperature values. Plotting the values in terms of anomalies leaves the impression that we are talking about big changes when physically that is not so.

The top line is also not dramatic if you understand what sorts of historical temperature statistics other scientists have produced. Readers having no familiarity with the field called paleoclimatology might get the false impression that the hockey stick graph is the only one on offer from

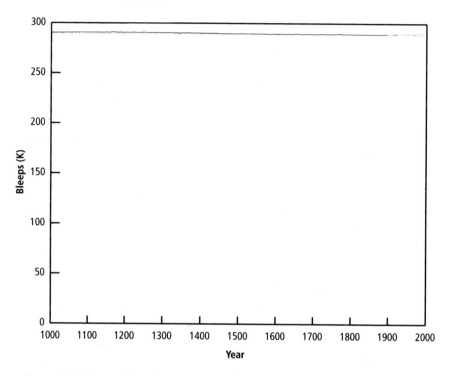

Figure 5.5. Figure 5.4 redrawn on a Kelvin temperature scale.

the expert literature. Far from it! It just happens to be the only one the Big Panel selected. They might instead have chosen Figure 5.6, which was published about a year earlier in the same journal where the hockey stick was published. It shows a series as drawn by some geologists who took a stab at reproducing past temperatures by measuring how temperatures vary at depths down thousands of boreholes in the earth's surface.[3] They argue that the vertical pattern of temperatures reveals the historical sequence of temperatures at the surface. As with tree ring proxies, not everyone buys this argument, but the authors at least have data from over 6,000 sites on every continent around the world, and there is as much reason to take their line as seriously as the hockey stick curve. But it does not show a smoking gun; instead, it shows something remarkably like the top line from Figure 5.4. You will not see this curve anywhere in the SPAM or any other Big Panel report. Apparently they were only looking for a hockey stick.

We are not making any judgement about whether the line or the

hockey stick is a more reliable picture of the history of T-Rex. Let specialists in the field debate that if they like. Both views have their strengths and
weaknesses. The point, however, is simply that there is a debate, because
temperatures over the last 1,000 years are not obvious to specialists in the
field. Yet a reader of the Big Panel's SPAM will learn nothing of this
debate. The caption under the hockey stick curve in the SPAM reads simply: "The year by year (blue curve) and 50-year average (black curve)
variations of the average surface temperature of the northern hemisphere
for the past 1,000 years have been reconstructed from 'proxy' data cali-
brated against thermometer data." That's it. The report itself adds some
cryptic hints that there are continuing debates going on, but nothing that
would mean very much to someone who did not already know the whole
story. Figure 5.1 is simply a work of mathematical artistry on unlike things
and is unsuitable to base policy on. The Doctrine strikes again.

Another idiosyncrasy of the field, where proxies are employed, is that
they would call Figure 5.1 a reconstruction. The people in the field know
that it is not really a rebuilding or a restoration of a former structure, as

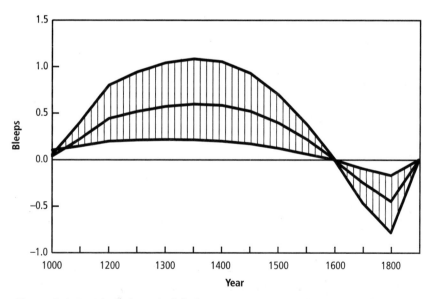

Figure 5.6. Borehole-based global temperature statistic, AD 1000 to 2000.
The horizontal zero line represents the present-day temperature statistic.
The darker middle line is the central estimate. The upper and lower lines
reflect probability bounds using Bayesian methods.

"reconstruction" suggests. You cannot have a replica of something that never existed. Calling Figure 5.1 a reconstruction is another instance of the simulacrum fallacy. Instead, it is properly the output of a model. It is *con*struction, a totally new thing, even if it purportedly tells us about things in the past.

Insiders in all scientific fields contend with odd sorts of linguistic conventions all the time without being fooled. None of them would say, for example, an anomaly other than zero is anomalous. The language means the opposite of what it sounds like. In statistical problems, differences from the mean are normal and not unusual at all, even if they are described as anomalies. You hear a similar example all the time on television weather programs. The commentators often discuss whether today's temperature was above or below "normal." It would be highly unusual if the temperature were "normal" for long. It's all just perverse jargon.

Specialized fields in science swim in a soup of such jargon. If you don't know the language, you can get the wrong—even opposite—impression. Doctrine supporters mine plots like Figure 5.1 out of their professional context, in which people know how to mentally translate the jargon into properly qualified terms. People who do not know that jargon, and do not know the conventions, can easily take the wrong impression from a graph like the hockey stick.

If you don't know that the numbers are constructed from a model, *not* reconstructed; if you don't know about the modelling convention of stationarity; if you see that the model is said to produce "anomalies" and don't understand that this means *normal* variability; if you are unclear about the difference between temperature and a statistic contrived to have temperature units, and you do not understand that temperature is only one from many properties of climate; if you do not know that measuring temperatures at airports to represent climate at large is problematic, and you do not know that the places where temperatures are measured has been systematically declining in many places; if you do not know how minuscule the anomalies are in terms of what the physical formulas take in; if you do not know that the mapping used can generate any shape of hockey stick you like, with no way to choose among them; then the hockey stick looks like a smoking-gun proof of a global warming problem.

Furthermore, if you imagine that statistical methods will warn you that you are constructing a model that produces fine-looking nonsense, think again. If you do not know these things, and you do not know how to ask the right questions to find out, you will do what bureaucrats and cabinet ministers and scientists and journalists and hundreds of others have done over the past couple of years: You will stare in wonder and think the climate has mutated and become a hideous beast we must wrestle to the ground before it destroys us.

That said, the research work behind the hockey stick is quite interesting. As a scientific work, it attempts to produce temperature fields from the past in ways with methodological significance. As a scientific work, it is fine. It does not need to be right, and it does not need to satisfy any other criterion than that it interests other scientists and inspires discussion and debate in a manner appropriate to the state of the art. The notion of an average temperature statistic is a stylistic calculation that is in current demand. The authors have provided one because they can. Physically, it means little. The hockey stick was almost an afterthought. Who could have known that it would become the poster child of the Doctrine?

The Hockey Game

The hockey stick cannot be the basis of policy. Extracted from its scientific context of incompleteness, trial and error, it means little. The Earth has not experienced the warmest global temperature in a thousand years: No coloured plot can support the claim that it has. And it is not a particularly relevant claim in any case.

Some might say that the hockey stick is remarkably abrupt, so maybe there has to be something to it after all. Or maybe there really is some little connection between global statistics and local proxies, despite all of the arguments to the contrary. For those people, we have chosen a different approach to persuade them how non-smoking-gun reconstruction methods are. Hockey sticks are really not that remarkable. You can find them nearly everywhere with this sort of technique. They are even in the most unlikely and meaningless places.

We have made a "reconstruction" of the U.S. real gross domestic product (a measure of total economic activity) back to the year 1000 using published tree ring data from Mount Washington. The result is plotted in Figure 5.7.

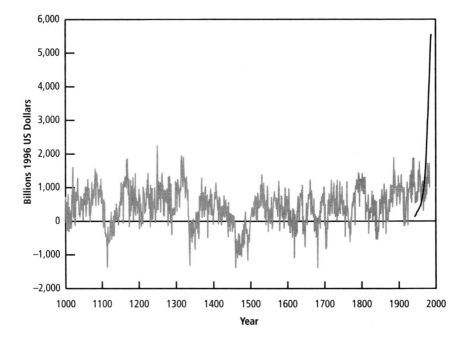

Figure 5.7. U.S. real gross domestic product, AD 1000 to 1998, using tree ring proxy-based mapping for the first 946 years.

We thought this particular monster, wherein two positively unlike things were so easily stitched together to get a pretty good hockey stick, would be really convincing to people. We thought this would make them much more cautious about interpreting such mappings and reconstructions in future. However, some comments from people were along the lines of, "Well...actually... there was an effective GDP before 1900, even before it was recorded. Maybe tree rings are picking up some of that." There seems to be little concern that the reconstructed GDP became negative from time to time.

We tried another reconstruction, this one of the German consumer price index, using the Mount Washington tree rings again. There was some hope that no one would speculate about it existing before it was recorded, and Mount Washington is quite a bit farther from Germany. Our reconstruction is plotted in Figure 5.8.

Here again we find a wonderful hockey stick. Incidentally, the statistics tells us that the mappings for Figures 5.7 and 5.8 "explain" about 13%

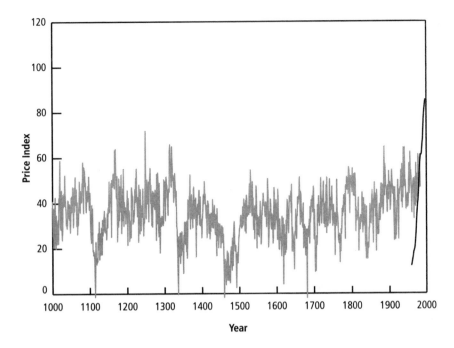

Figure 5.8. German consumer price index, AD 1000 to 1998, using a tree ring–based proxy for the first 959 years.

and 24% of the target values, respectively. Not bad compared to the temperature hockey stick!

Hockey sticks are everywhere in the reconstruction business and are not remarkable in any way. The dramatic appearance is purely one of joining a fairly steady signal onto one that is systematically rising. It does not mean that you have discovered anything at all about the past based solely on the statistical method. It may mean you have only constructed, as we have, patchwork monsters using a technique worthy of Dr. Frankenstein.

---

[1] Data source: Mann, Michael E., Raymond S. Bradley and Malcolm K. Hughes. "Northern Hemisphere Temperatures During the Past Millennium: Inferences, Uncertainties and Limitations." *Geophysical Research Letters* 26, 1999, 759–62.

2   Mann, Michael E., Raymond S. Bradley and Malcolm K. Hughes. "Global-scale Temperature Patterns and Climate Forcings over the Past Six Centuries." *Nature* 392, 1998, 779–87.

3   Huang, Shaopeng, Henry N. Pollack and Po Yu Shen. "Late Quaternary Temperature Changes Seen in Worldwide Continental Heat Flow Measurements." *Geophysical Research Letters* 24, 1997, 1947–50. We are grateful to Shaopeng Huang, who supplied his data for Figure 5.6.

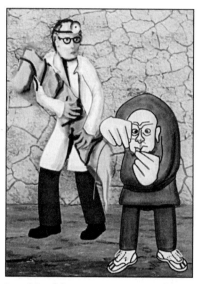

Igor! I told you to put away the needle and thread! We're using a statistical methodolgy to attach the fresh limbs.

6 The Unusual Suspects

Climate and Punishment

In the last chapter, we told you that the hockey stick curve is not a smoking-gun proof of climate change. This raises the whole question of how one would detect climate change in general and the human influence in particular. The Big Panel calls this problem signal detection. In one sense, they are more right than they realize.

The jargon is inherited from old radio technology, where there is a carrier wave and a signal plus noise. A carrier wave has a very simple shape so the receiver knows what to look for. The signal rides piggyback-style on it. Sometimes additional unwanted fluctuations appear, called noise. Statistical methods are sometimes helpful for separating out the noise to make the signal clearer. Radios and televisions work because we know how to separate a signal from the carrier and noise.

In recent years, people have figured out how to carry a signal on a *chaotic* carrier instead of the simple wave of old radio. Chaos was discussed in Chapter 3 as a property of unpredictability in solutions emerging from certain systems of nonlinear differential equations. We cannot solve the associated differential equations directly. The solutions we get by other means appear as noise, even though they really are not noise.

Why would someone want to send a signal on a chaotic radio wave? The answer is simple: So that no one other than the person for whom the message is intended can understand the message. This is the essence of the subject of cryptography. In the case of the chaotic carrier, it looks so much like natural noise that you would be hard pressed to recognize that there is any signal to find at all. If you do not know how to generate precisely the right chaotic carrier, it can be very difficult to figure out what the message is and accordingly get at the signal. With this technique, you can even send encrypted voice messages in real time.

This brings us back to global warming. The "global warming signal" is being carried, not on a simple wave or cycle, but on a naturally chaotic climate carrier. We do not know what the carrier is, for reasons we have outlined already. It is a form of natural cryptography. But people think that they can figure out the signal anyway. They think that classical statistical methods will work. Just let them try to crack the far *less* difficult chaos cryptography systems in existence. It would be interesting to see how far they actually get.

The other sense of the term "detection" is what Lord Peter Wimsey meant when he said he "dabbles in detection," namely, that he works as a detective. He figures out whodunnit. The signal detection literature reads a bit like a spy novel, with all its code-breaking and signal detection language, but it also reads like a murder mystery. There is a lineup of suspects, teams of sharp-eyed sleuths, interrogation, fingerprints, degenerates who won't confess and a grave sense that the guilty must be punished.

The Big Panel sometimes talks about two separate stages: detection and attribution. But they are not really separate steps at all. The former refers to gathering evidence that the climate has changed. The latter refers to finding a causal connection between such climate change and fossil fuel use. The existence of climate change itself is not something in need of detection. Putting out a 1,000-page report announcing that the climate has changed in some sense over the past century is as pointless as putting out a one-page report announcing that time has gone by over the past century. The Earth's condition is, after all, the result of a chaotic dynamical process. It would be remarkable if there was something that had not changed on century time scales, even in the absence of any exterior causes.

Instead, the PUN wants to be able to say that the climate has changed in some *remarkable* way so as to thicken the "detection" plot. They are not so interested in "natural" climate change (i.e., the carrier), they are looking for some big "unnatural" climate change (i.e., the signal). But detecting unnatural or unusual climate change is what some thought the attribution step was supposed to be about, because you can only cause something that wasn't going to happen anyway.

You can only have attribution if "unnatural" climate change has already been detected. But any detection that merits more than a one-paragraph

PUN report, presupposes something that is "unnatural" and by implication something to attribute it to.

While we say that we expect climate change and that it would be surprising if anything had not changed, it remains difficult to find evidence of remarkable or unusual climate change. Moreover, it is far from clear what would constitute evidence.

There is no single definition of "climate," nor a theory to tell us what its behaviour over the past century would be in the absence of fossil fuel use, nor is there even physical guidance on how to measure and characterize its current state.

Most attention is fastened hard on the surface-measured "global temperatures" as in Figure 4.1 (T-Rex). There is a bit of an upward slope. But maybe the "natural" progression of that particular average would have been upward at twice the rate, and carbon dioxide is actually slowing it down. It is certainly possible. Of course people who always think in terms of "average heat" (let's call them "average-heat theorists") will have difficulty imagining this. Anyone who forgets that clouds, wind and local state changes, as well as the vagaries of turbulence, cannot be rolled into some arbitrary average to represent some non-physical "total heat" in the system will assume this cannot happen, but it can.

The fact that weather satellites pick up no averaged warming in the troposphere (see Figure 4.3) is often cited as evidence against "global warming." But how do we know what that particular statistic should "normally" look like? Maybe we would have experienced cooling in the atmosphere and carbon dioxide is causing it to warm. The laws of nature do not limit the Earth's climate to some static state in perpetuity.

However, the Big Panel works on the explicit assumption that the climate should be stable and unchanging in our era, according to some statistical measures. You might rightly commend them in being explicit about it, but this really is a huge assumption. At least it solves the problem of changes in the carrier being confused with something that could be "attributed" to something unnatural. With this assumption, any climate change is "unnatural" by definition.

The Big Panel has a diagram (Figure 14 in the Technical Summary) that shows flat but wiggly lines drawn by some climate models. These are said to be computer-generated pictures of what T-Rex would look like if people didn't assault it with infrared gases. Then they have pictures of a particular T-Rex based on measures from white boxes over a period when there was a lot of carbon dioxide emissions. That T-Rex slopes up. One conclusion might be that CO_2 does not matter, but the models are no good at simulating T-Rex. Another possibility, and the one that is implied strongly in the report, is that the models are quite accurate, infrared-absorbing gases did violence to T-Rex, and ipso facto unnatural climate change has been detected.

We will not try to demonstrate which conclusion is the truer one. The example serves to highlight, though, that signal detectives are working under the disadvantage of trying to solve a crime while not knowing if a crime has even been committed.

In an ordinary detective story, there would be a scene where the medical examiner shows the detectives a gunshot wound or tell-tale strangulation marks and confidently explains why the victim could not have died from natural causes. In the climate detective story, there are no gunshot wounds or strangulation marks. Nevertheless, in their big report of last year, the Big Panel ruled out natural causes and concluded that the world has warmed and that most of that warming over the past 50 years is attributable to human activities.

One of the great plot lines for a mystery is where some rough lout stands convicted of an awful crime, but someone else thinks he is innocent. The gavel comes down and the judge passes a long sentence. At that very moment, a hard-boiled private eye gets a gut feeling the cops nabbed the wrong man.

He does a little digging and finds out that the constable first on the crime scene found no sign of foul play and assumed the victim had suffered a heart attack. Nor did the coroner find any wounds on the body. How could the arresting officer be so sure it was violent homicide when it was not even clear a crime had occurred?

Then the detective discovers that the lineup was rigged to lead the eyewitness to pick the accused. And that there were many people seen in the vicinity of where the victim died, not just the accused—but for some reason the other suspects were ruled out without being asked for alibis. And the accused had two large black eyes and a broken hand after the interrogation at which he allegedly confessed. Gradually the intrepid gumshoe begins to peel away the layers and show that the investigation and trial were railroaded.

You can write the rest of that script. The script we will write here begins with the same question. How is it, if we cannot define the climate or characterize its natural behaviour, that can we nevertheless say an unnatural climatic change has occurred and that humans are to blame? We will go over the case step by step. We will tell you about the narrow field of suspects the Big Panel looked at, how they were interrogated, who was not investigated and how the authorities got a confession out of carbon dioxide.

There are a million stories in the naked climate. This is one of them.

The Lineup of the Usual Suspects

The story opens with Figure 3 in the SPAM, which we have reproduced as Figure 6.1. It is a sort of police lineup of the suspects. The crime under investigation is global warming, and these suspects are shuffling in because they are all under suspicion of interfering with something called the "radiative forcing of the climate system."

Radiative forcing is one of those concepts that looks like a simple physical variable but turns out to be the product of a model that takes physical

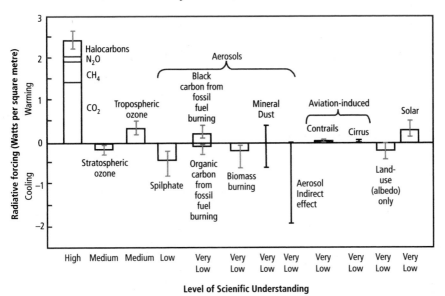

Figure 6.1. The lineup of suspects.

inputs and produces strange and unintelligible numbers as an output with all kinds of hidden, built-in assumptions and provisions that are not clearly reported. As with T-Rex, it involves taking averages of thermodynamic variables that are not meant to be averaged together in order to reduce the climate to a one-dimensional concept like "warming" or "cooling." And, like the hockey stick, it involves mapping unlike things onto each other so they can be compared on some axis.

The stated units of measurement are watts per square metre at the surface. Two watts per square metre sounds like something you would get if you shone a nightlight over a patch of grass—and in fact Big Panel people sometimes use that very analogy to describe what infrared gases do.

A watt is a unit of power, which is just energy used per unit of time. Everyone seems to understand what energy is these days, but they are puzzled when it is used per unit of time. They think it is electricity or something. It has nothing especially to do with electricity, despite what at least one edition of Webster's dictionary says. The unit of energy here is the joule, and a watt has units of a joule per second. Electrical products like heaters and stereos and light bulbs and hair dryers are rated in watts. So are cars (in Europe). In North America, we use horsepower for cars and lawn mowers, unless they are electric (there we go again). One horsepower is 746 watts. So a 100-horsepower car is a 7.46-kilowatt car.

We can figure out how many horsepower are required to operate a 75-watt light bulb too (not a standard size, but you can get them, and it makes the example simpler). In that case, slightly over one tenth of a horsepower is required. Ten of those bulbs would require about one horsepower.

If you took 100 times more, or 1,000 light bulbs in total, and stuck them into a bank of lights, you would need to use about the same amount of power as the car. It would produce a lot of light. But it doesn't mean that if you shone all 7.5 kilowatts on the car, it would drive away. Nor could the sound from 250 stereos, each rated at 300 watts, make your car drive away, no matter what you are playing.

Could we say that 250 × 300 watt stereos are as loud as 1000 × 75 watt bulbs are bright? No, of course not! Stereos make sound, not light, and light bulbs make light, not sound, at least when they are working right. The comparisons are meaningless. Oh, we could get into details about how to trap the sound and have it do some sort of work to generate electricity, or we could consider how photocells can take the light and convert it into a form the stereos can use, and we could roll in some mechanism to account for losses of power on the way too.

Still, these considerations establish that you need a *mechanism* that links the energy to the process you want to happen. It is not good enough to imagine that you have a big pot of joules that we let out at some rate and things just happen. You need a specific device to do a particular thing. One device cannot be exchanged for just any other, unless you bring in a new device to do a conversion. The key is the word "used" in the definition of power. It implies a particular mechanism in the realization of power.

So what are the mechanisms that use these watts per square metre in Figure 6.1? By rendering all these climate variables in watts per square metre, two illusions are created. The first is that the mechanisms don't matter. How else would they be in the same lineup otherwise? If they did matter, it might be like comparing loudness to brightness.

Is the effect of each of these things on climate the same? Does one watt per square metre of one have the same effect as one watt per square metre of another? This is far from obvious, especially since we are not sure how to physically define "climate" in the first place.

The other illusion is implicit, generated by the premise that we are talking about global warming and that we may equate it with heat. Average-heat theorists, who embrace this as part of their **ambient heat prejudice**, might be able to get away with it if the atmosphere were as hard and transparent as Lucite and we were trapped like flies in amber. Of course there would be no climate to think about then. But alas it is not that simple, as we have taken pains to point out. The medium of energy exchange in the real world moves when it receives energy in particular ways, and it changes its local condition and state too. It doesn't just warm up, any more than light bulbs produce music.

They might say it warms up on "the" average. So their idea is that T-Rex is a sort of "Lucite approximation." After the last three chapters, it should be clear that such a Lucite approximation has little to do with Lucite, as if we cared what Lucite was doing anyway. But that is what was on their minds. Indeed, when this sort of lineup was presented in former times, long ago, the units on the vertical axis would typically have been changes in temperature instead of watts per square metre. The heat theorists are not so obvious anymore in their bias toward always thinking in terms of heat, a propensity we will call *the heat prejudice*.

The first warning that there is something peculiar about this figure comes from your knowledge of Chapter 3. Note that the most important infrared ("greenhouse") gas is not there! It is astonishing to present a list of infrared-absorbing gases while leaving water vapour off the list. It is like describing the PC software industry and leaving out Microsoft.

Why is it left off? The reply will be that it does not "force" climate but is a property of it instead. But this is an artificial way of looking at things. It presumes that "forcing" is only done from the exterior, and internal dynamical processes do not freely shift into different regimes, forcing other internal processes to change. It is a consequence of the quiescent climate assumption introduced at the outset, which allows for no variation in the climate in the absence of exterior forcing.

Are these values in watts-per-square-metre measurements of change in the item specified, or do they include the contribution from changes in water vapour too? That is important because of how water vapour boosts all of the sensitivities. In the jargon of climate research, there is a "positive feedback" from it. In the model, a bit of some "forcer" increasing leads to slightly higher temperatures, which lead to more water in the atmosphere. The powerful water vapour absorption then takes over.

The whole effect is piggybacked on that powerful absorption. So if they do not include the effect of water in the numbers, then they are artificial. If they do, then they have large built-in assumptions about how water vapour changes thermodynamically and dynamically in response to exterior forcing, which involves all of the fatal scientific problems from Chapter 3. This draws into question the claim that the column on the left reflects a high level of scientific understanding.

Furthermore, if we do not use the hidden model assumptions, there is no basis to compare, say, aerosols to sunlight. We have to face the fact that these very different things are associated with very different mechanisms, which render comparison of power not unlike stereo watts to light bulb watts. We can only compare over what these things do rather than some contrived power numbers.

Aerosols scatter sunlight, but do not produce it. The sun produces sunlight, but does not produce aerosols. Infrared absorbing gases do neither; instead, they capture and re-emit bands of radiation. Land use changes affect local reflectivity, which is something different altogether. These effects are not simply reducible to energy rates at the surface, nor are they measured in watts in any context in which their effects make physical sense. The only way to take these mechanisms and convert what they do into energy per unit area is with a device. In this case, the device is the climate mechanism. We don't know how that works, so we use a model instead.

So these watts are not measurements or computations from an exact theory. They are numbers from models, which have built-in assumptions about what is included. One of the most important assumptions is what gets left out. If you do comparisons on the basis of the model assumptions, the result can be wrong. But if there are mechanisms *not* included that really do exist, then the comparisons can be *very* wrong.

For example, it doesn't take a lot of energy to start 75 kilowatts of car. All you need to do is to turn the key. The meagre energy required to turn the key will do what 1,000 light bulbs could not do. We will call this the amazing key-turning effect. A mere turn of the wrist unleashes 75,000 watts of power.

In a climate context, it is easy to find examples of mechanisms in which key-turning effects are likely to happen. Consider sunlight, which is not heat, despite what some people seem to think. It can be converted to heat on absorption, but how that takes place is the issue. Sunlight is a collection of photons with different energies. When the energy of the sun varies, it does not vary uniformly over all photon energies. Its brightness varies enormously in the higher-energy photons, for example, and these changes could lead to changes in stratospheric chemistry, which could lead to other changes in dynamics. It is not the total energy that would make the dynamics different; it is just that it sets up the circumstances to redirect other energies. If you think of sunlight as heat, such a scenario cannot be seen.

We are not saying that a mechanism like that has been proven. We are saying that no one really knows for sure. It is no use trying to say that it is simpler to assume that no such mechanisms exist. Thinking that way doesn't work in this problem. The correct answer is that no one knows. Besides, nearly all of the processes that the PUN is concerned with are small things that get amplified through a mechanism. The water vapour feedback mechanism is an example. It would make sense, then, to think the issue is about mechanisms and not simply energy.

Putting the solar change on the lineup makes it seem like people do know something that they do not. We will talk more about that in a bit.

Meanwhile, this key-turning effect means that not all watts are equal in a complex physical process integrating many distinct physical mechanisms. Given the amazing key-turning effect, the hidden model assumptions and the ambient heat prejudice, we are left wondering what "watts" actually mean in Figure 6.1. The PUN authors have crafted an illusion of comparability between very different things. In 6.1, the numbers are not from measurements, and the sizes really depend on the hidden mechanisms of the model definitions.

Accordingly, for the lineup in Figure 6.1, where the model mechanisms by which diverse phenomena convert energies to power units are unknown and maybe unsound, we will use "whats" instead of "watts." As before, you can always do the conversion back. But it must be done consciously, making it difficult for those easy backdoor misunderstandings to creep in. In this language, these models map the diverse things listed—the atmospheric pressure of CO_2 (in parts per million), the reflectivity of land, the radiation spectrum from the sun, etc.—into whats.

Since the physical principles are not well understood and the relationships being modelled change constantly as local atmospheric conditions change, the "whattage" associated with each of these things has to be averaged in various ways. We already explained (in Chapter 3) why we would not expect to be able to do such a grand average calculation. This is true for infrared gases too as a consequence.

Yet by comparison to the others, the Big Panel acknowledges that even this uncertainty is nothing compared to that for the other mugs in the lineup. According to the "Level of Scientific Understanding" labels along the bottom, the infrared absorption problem rates a High while the rest taper quickly off to Very Low. It is hard to imagine how the level of understanding could be lower than it is for the infrared gases, but that's what the picture says.

The presentation of Figure 6.1 invites you to look over these characters to see if anyone looks guilty of messing up the climate. Unlike things are averaged and mapped into whats, using some models you don't get to inspect. Most witnesses will not think about these things, though. Instead, they will finger that big guy on the left, namely, the infrared gases. He kind of stands out. Then again, the lineup was set up so he would. It's as if the police put a man suspected of a bank robbery in a lineup with a bunch of toddlers.

By expressing everything in common terms of whats per square metre, you would think we could add these things up to get a "total" forcing effect. You can add energy per time, can't you, Professor Thermos? But these numbers do not work that way and Professor Thermos would not approve. It is like adding the output of the stereo to a light bulb in watts. Indeed, the text in the SPAM underneath this picture rightly warns against trying to add them up.

But if this is so, what are we actually comparing in Figure 6.1? In one of many acts of self-contradiction, they went ahead and did just that by adding up the infrared gases in one column! It makes as little sense as adding aerosols and solar flux. The infrared gases are active at different altitudes at different photon energies. Changing one has different model implications than changing another. And how do you determine comparable changes in one in terms of another? Their influence on the climate cannot be easily defined independently of each other or of the local conditions.

One specific example of why these "forcing" suspects cannot be added together is that as any one is added to the atmosphere, it changes the distribution of water vapour, which affects all the other factors. For instance, additional water vapour shortens the length of time molecules of ozone and methane stay in the air before raining out, depending on model water cycles.

The bars that are shown are just model conjectures, although such bars are usually placed on known ranges of errors. The little error whiskers (the I) are not statistical or physical computations. They just reflect the range of numbers in some published papers the Big Panel happened to read. Had they read more papers, the ranges would have been even wider. And behind them are models that could produce an even wider range of outputs, if you wanted them to.

We will explain in the next chapter why the bar for carbon dioxide, notwithstanding its height here, could be very different. The real degree of certainty associated with the infrared gases is quite low—and as the figure says, it is lower still for all the others.

The rogues' gallery of "radiative forcing" characters presents a seriously incomplete list of suspects, transformed by unknown mappings and averaging models onto some physically meaningless axis, based on very low levels of scientific understanding. The amazing key-turning effect is not considered and the ambient heat prejudice is rampant. In the PUN court, this was enough to secure a conviction. No wonder that private eye is having doubts.

Who are the characters in the lineup besides the infrared gang? The aerosol brothers are well known in environmental crime circles, while the little solar punk over on the far end seems pretty harmless. However, now that they have been seen in the company of the infrared gases, they have attracted a lot of suspicion. We will tell you a bit about the rumours and investigations before talking about another big puzzle of Figure 6.1: the long list of shady characters who for some reason were not included.

The Aerosol Brothers

By the mid-1990s, comparisons between actual and predicted changes in the temperature field over the 20th century showed that the main climate models used by the PUN exhibited far too much sensitivity to CO_2 growth. In the jargon of the field, they "ran hot." If infrared absorption by CO_2 and other gases caused as much warming as GCMs are programmed with, our world would be a noticeably warmer place today than it actually is.

Therefore, if you wanted the models to mimic the real world, the GCMs would have to be changed internally or new effects introduced. Of course the tendency was to tune the models to the surface T-Rex and not the average from satellites. The ambient heat prejudice was at work here too. Rather than drawing down the CO_2 sensitivity or reworking the fluid dynamics to introduce negative feedbacks, many researchers proposed instead that aerosols, in particular those associated with sulphur dioxide emissions, exert an offsetting (cooling) effect.

From the beginning of the idea of global warming, there was always a battle between aerosols and infrared gases. Would it be cooling or would it be warming? The answer would always depend on things yet to be known. Returning to aerosols to fix "running hot" was like going home. It was the inevitable place to turn based on the history.

Aerosols are very small particles of matter that arise from many sources: sea spray, dust, pollen, etc. Aerosols can be formed from every known substance. As we explained in Chapter 3, the role of aerosols in atmospheric fluid dynamics is deeply complex, and no one is in a position to make any specific claims about exactly how they affect local atmospheric temperatures.

Attention has gotten stuck on sulphur. Most fossil fuels contain some of it, and when they are burned, the sulphur adds to the stock of aerosol particles in the atmosphere. They may scatter some sunlight back to space. By providing nuclei around which water droplets form, they may also make clouds longer lasting, although exactly how, quantitatively speaking, is hard to say.

Aerosols caught up in a turbulent flow change the dynamical and thermodynamical conditions on a microscopic scale, which in turn changes where and when the wind blows and whether energy will manifest as latent heat, sensible heat or some other form. All these things matter if we want to connect aerosol loads to climatic influence, but none of them can be treated directly in a global climate model.

Aerosols were hypothesized to exert a strong enough cooling effect on the whole climate system to offset the influence of infrared gas buildup. If true, this could explain why CO_2 emissions in the 20th century generated a smaller increase in T-Rex than expected. And since SO_2 emissions and atmospheric concentrations have been reduced since the 1980s (through policies to control acid rain), their effect is being lost. Hence the postwar pattern of T-Rex falling, then rising, is broadly consistent with a sulphate+CO_2 hypothesis.

But "broadly consistent," in the absence of any theory or compelling empirical evidence, is not enough to secure a conviction. It is circumstantial evidence. We prefer causality, which requires an understanding of the physics. Otherwise, the best we can do is look for correlations.

Even then, the only things that can be correlated are indexes of the effects of "CO_2+sulphates" with T-Rex. Both of these are products of climate models, not direct observational data (we will discuss this in more detail below). Moreover, while emissions of SO_2 from fossil fuels have been estimated back to the late 19th century, atmospheric concentrations (including those from volcanic and oceanic sources) are not available that far back.

As you can see in the diagram, the Big Panel guesstimates that the indirect local "radiative forcing" effect of aerosols ranges from about 0 to −2 whats per square metre. It is not pejorative to call this a guesstimate. The indirect cooling effect of aerosols is represented *only* by a long "whisker" in the diagram.

Despite having no clue, a *strong* indirect aerosol cooling effect is programmed into the models used for Big Panel simulations so as to generate a broad cooling effect on the global climate. This assumption is essential to maintaining the existing model effect for CO_2. Even ignoring possible heat biases due to urbanization and the loss of measuring stations, without a strong indirect but shrinking cooling effect due to sulphate aerosols, the models substantially overpredict warming. An open-ended indirect sulphate cooling effect serves as a big fudge factor.

Furthermore, the models would overpredict even more if the satellite temperature statistic was used to tune them. That would in turn take more aerosols.

One big difficulty with assuming such an effect is that burning fossil fuels releases a mix of aerosols, not just sulphates, in particular, black carbonaceous soot. The bar for black carbon aerosols is very small in Figure 6.1. It is as if the PUN put them in the lineup and hung a big sign over them saying "Not these guys."

Yet after that diagram was prepared, some results appeared in the scientific literature that suggests the black aerosols are more trouble than previously thought. We now know that soot absorbs infrared and produces a measurable local warming effect. And one of the things learned recently is that atmospheric soot goes through a chemical reaction after release that binds it to other aerosols, including sulphates.[1] Once mixed, the soot molecules have shown a local warming potential in this model context that nearly balances the cooling effect of sulphates.

These findings about the role of soot are important. PUN detectives ignored soot emissions. In fact, the literature on which the Big Panel based its conclusions concerning infrared gas warming effects assumes sulphates do a lot of cooling and soot does nothing. Yet soot appears to be exceeded in importance only by water vapour and CO_2 in terms of warming effect.

Including them in signal detection studies would force down the warming attributable to CO_2. Even more significantly, it would challenge the assumption that aerosols as a group have a strong cooling effect. (That was never more than a guess anyway, and now it just seems an unlucky one.) If sulphate aerosols are not cooling the atmosphere, and soot is contributing more warming than previously thought, then the climate sensitivity to CO_2 as programmed into GCMs is too high.

It is worth remembering that the sensitivity of the climate to CO_2 is an input to the model. Someone has to pick a parameter and type it in. It is not something the model predicts. If this parameter is wrong, the whole modelling exercise is bound to go astray. But these kinds of bad deals happen when you hang out with bad company such as models.

The Solar Punk

That bright orange thing in the sky is a ball of turbulent plasma riddled with enormously powerful magnetic fields that reach out far beyond us, beyond the outer reaches of the solar system.

It is a star—a very unusual star. Since we have observed it with modern techniques, it has exhibited behaviour that is extraordinarily steady in comparison to other stars of its class.

When most people look at the stars, they presume that behind the twinkling not too much is going on and the stars are steady beacons in the night. Nothing could be further from the truth. There is a star called Algol in the constellation Perseus that you can watch over the course of a few days. You can see it for yourself with your naked eye. It is was known as the demon star because in ancient times when its behaviour was discovered, stars were thought to have a fixed brightness. But Algol changes in energy by more than a whole stellar magnitude, which means that the energy from that star varies by a factor of two to three times during the few days of its period.

There would not be much scope for life as we know it with that sort of change in our star. Algol changes in brightness because there are really two stars eclipsing each other. Obviously we have only one star where we live. That's a good thing. But there are many other lone stars in the sky that change noticeably over any time scale and brightness scale that you can imagine.

Nearly all stars vary in brightness to some degree, but there are some classical types. There are the Cepheid variables, for example, which were instrumental in getting at the scale of the universe, because their brightness varies with their period of oscillation.

Other types have names like surnames with initials. There are RR Lyrae and RV Tauri and U Geminorum stars, to name some. The last of these types is pretty steady most of the time, except every year or so it erupts, with 100 times as much power being radiated as normal. This lasts for a few days and it then returns to normal.

R Corona Borealis stars are the opposite. They run along steadily and then suddenly drop in brightness by 10,000 times and gradually climb back to their original steady condition. In the beautiful Orion Nebula, there are newborn stars that you can watch vary irregularly in brightness with your own eyes over just a few hours in a small telescope. There are even stars that blow up and then seem to re-form. And there are many more wonders.

At the stable end of the range are stars like our sun. Among those, it seems that the sun, as observed over recent decades, is exceptionally steady. There is reason to expect that it was not this way in the past and that it will not be that way in the future. But the extremely small amounts of variation in terms of the total are why the PUN downplays the significance of the sun in changing climate.

But even if changes in the flow of energy to us from the sun were always so small, the sun almost surely influences the Earth in important ways. It doesn't just provide energy to feed the supporters of the heat prejudice; it also has much capacity for that amazing key-turning effect we defined at the beginning of this chapter. It generates a full spectrum of radiation and streams of energetic charged particles called the solar wind. Then there is its huge magnetic field that we are immersed in. The interactions with the atmosphere are surely far more complex than just causing heating. The issue of mechanism is every bit as important at the simpler matter of energy change.

Over the past century, it seems the sun got a bit brighter and the spectrum changed. As you can see from Figure 6.1, the PUN figures this had little effect on the climate, which in this case is summarized as a simple whattage at the surface. But they also admit to having little understanding of how the sun influences the climate.

One of the points of controversy has been what constitutes the most relevant measure of solar activity. Sunspot counts and measures of total flux have a relatively weak correlation with temperature statistics, so authors who use this measure conclude that the sun's role is very modest over the century as a whole, and especially modest in the last two decades. One study we saw tossed the sunspot count into a correlation model, found it did not match one particular T-Rex very well, and confidently concluded that the sun does not cause climate change.

Others have argued that the sun has a bigger influence, depending on how solar output is measured. In particular the *solar cycle length* is tightly correlated with T-Rex. There is an irregular cycle of about 11 years' duration during which solar flux slowly grows, then declines.

Some Danish astronomers found not long ago that averaged northern hemisphere temperatures from 1851 to 1987 follow the inverse of the length of the solar cycle with an impressive 95% correlation.[2] This kind of study has also been extended back to the 1600s using the long temperature series recorded at the Armagh observatory in Ireland. No one really knows what to make of this, but it is interesting, and it cries out for research into key-turning mechanisms.

Other research teams have looked at mechanisms by which the climate amplifies fluctuations in solar energy. One group at the Scripps Institute of Oceanography in San Diego has noted that upper-ocean temperature is correlated with changes in solar energy, but the total energy changes in the ocean seem to be between two and three times as large as the change in solar flux.

They hypothesize that variations in trade wind intensity explain the difference. Decreases in solar flux reduce wind intensity, which slows down the transfer of heat out of the oceans. This sets up a climate oscillator, which works in phase with the solar cycle, amplifying its climate effect two- to threefold.

Another Danish astronomer, Henrik Svensmark, put an alternative theory of solar amplification forward. He based his idea on a long-hypothesized relationship between cosmic ray flux and cloud formation. If cosmic rays do enhance cloud formation, then increased solar flux will affect cloudiness. Solar wind suppresses cosmic ray entry into the atmosphere. As solar flux grows, solar wind grows and cosmic ray flux declines and fewer clouds form. Since (low) clouds have a net local cooling effect on climate, the reduction of cloud cover during periods of high solar output will amplify the direct effect of the solar cycle. It is a nice little key-turning effect, if it could be proven.

The most recent solar cycle peaked in late 2000 and was on its way down when, in October 2001, the sun went nuts and jumped to a new, unexpected peak. It stayed there for three months before settling back down to its normal descent toward the next solar minimum, which is expected sometime around 2007. The burst of flux coincided with the unusually warm conditions in the 2001–02 North American winter. But it might have been a coincidence: Like the sign on the lineup says, there is very low scientific understanding of any of this.

The Less Usual Suspects

The lack of water vapour is an astounding omission from Figure 6.1. But H_2O is not the only scoundrel who got a free pass when this lineup was being put together. We can think of a couple of other shady characters who also should have been brought in for questioning: the moon and internal dynamics.

We will describe them in this section. But those are just the big guys. We will not even claim a comprehensive list. As we said in Chapter 3, no one knows how small something has to be not to matter. Every seagull, butterfly and snoring dog belongs in the lineup too, in part because of the chaotic internal dynamics. Then we will look at the forensic evidence and ask why anyone is so sure a crime was even committed.

The Moon

The moon affects climate through variations in its tide-raising force. The stronger the lift it exerts, the more water is pulled up from deep layers, cooling the ocean surface and in turn the atmosphere. Ocean researchers have long known that tide-raising forces follow cyclical patterns. While cycles on decadal and century scales have been identified, there had been occasional suggestions dating back as far as 1914 that a strong 1,800-year cycle also affects tidal mixing. Recently, Keeling and Whorff[3] of the Scripps Institute of Oceanography assembled a long enough time series of tidal force measures to identify this cycle.

They found that shorter cycles of about 200 years are embedded within an 1,800-year cycle. A peak coinciding with the Little Ice Age is clearly identifiable. We are now in a long-term phase marked by a diminishing rate of ocean mixing and warming sea layers, which the authors argue will continue to develop for four more centuries. Eventually it will reverse and the cooling phase will again be at its strongest at around 3200 AD.

They predict that the climate is going to get warmer in the next few centuries—warmer than anything we have experienced in the past 2,000 years—due to natural causes, irrespective of infrared-gas emissions.

Meteorological Oscillations

We now know that there are a number of short-term cyclic patterns in the weather system called oscillations. Meteorologists have identified many of them in recent years, including the El Niño-Southern Oscillation (ENSO), the Pacific Decadal Oscillation (PDO), the North Atlantic Oscillation (NAO), the Interdecadal Pacific Ocillation (IPO), the Cold Ocean Warm Land Oscillation (COWL) and the Arctic Oscillation (AO).

Two of these oscillations (the PDO and the NAO) seem to flip between "cooling" and "warming" modes rather abruptly. Around 1977, both flipped to warm. This coincided with a circulation realignment in the Pacific, shifting a pool of cold water from the Aleutian Islands to the southeast. This event caused a step-like increase in temperature that has been observed in temperature records from around the North Pacific, as well as other places around the world.

This so-called Pacific Climate Shift had large impacts on fisheries, bird migrations, coral reefs and so on. For instance, catch rates of Alaskan pink and sockeye salmon jumped in 1977. It was studying this abrupt change in fish catches that led marine scientists to identify the Pacific Decadal Oscillation.

Recently, the Pacific Decadal Oscillation seems to have switched regimes to a negative mode, suggesting that we could be in for a decade or more of cooling in North America. These meteorological oscillations are not true oscillations in the simplest mathematical sense, but irregular signals with an identifiable swinging back and forth within a fairly small range of frequencies. They are better described as quasiperiodic where certain frequencies dominate. The sun's 11-year cycle has a similar character.

The danger in the "oscillation" language is that people will think that the atmosphere and ocean are just a bunch of distinct periods with separate processes piled up on top of each other. They certainly are not. Many of them may prove to be transient structures reflecting longer time-scale processes than we have been able to observe over the time scales we have been seriously looking at. Some of them may be due in part to external "forcing," either directly or by some "turn-key" mechanism, whereas others may have no external cause and just be the process doing its normal "breathing." We will talk more about that in the next section.

In any case, it is clear that these relatively long time-scale shifts between regimes draw into question any assumption by the PUN that the natural climate is steady on the scale of decades and more.

A Second Opinion on the Autopsy

In this section, we will revisit the question of how anyone knows a "crime" was committed, i.e., that some "unnatural" change has occurred in the climate. The case was stacked against the defendant by presuming that there were no natural dynamics other than some steady, classic background noise. There is an assumption that the dynamics of the climate are statistically very simple, even if they may not be.

Unless something forces it from outside the system, any particular average you construct (e.g., temperature, "radiative forcing," etc.) ought to be flat over time. It is all very cozy for the heat theorists. A simple random carrier providing a steady climate fits in well with this view.

Figure 14 of the Big Panel's Technical Summary works on this assumption. Since the models could not reproduce T-Rex when they are run on the assumption that humans are still in the trees, the conclusion is that humans have changed the climate. There is little sense that more work is needed on the models. Models are not perfect, they will say, in classic doublespeak, but they can't be wrong. They must conclude it is human moral turpitude that is the cause of the discrepancy. Discovery of a trend in T-Rex, or some correlation between particular averages and outside influences (like carbon dioxide concentrations in the air) is taken to mean that climate change is happening, and that is that.

The Hammer and the Poodle

There is a saying to the effect that if all you have is a hammer, everything looks like a nail. Classic statistical methods are the hammers and time series all look like classic stationary noise processes.

Figure 6.2 is a time series of 2,000 values. It could be an anomaly from any number of measured climate properties. Suppose it is carrying a signal. How would you know? Perhaps we could use statistical methods.

Normally, seeking a signal involves separating the data into the signal—the thing we are looking for—and a simple known carrier (e.g., Figure 6.3) and some true noise that we can subtract out and throw away. If done right, the noise is small. Radio technicians do this sort of thing. The signal rides the carrier.

Figure 6.2 looks like it is all noise without the classic carrier at all. But it does not have any noise in it at all. It is in fact a carrier that has successfully transmitted a signal. It is not a simple one as in classical radio, but it is a carrier all the same. This is remarkable because carriers for radio and television usually look more like Figure 6.3, which is anything but noisy looking. However, people have used data like that in Figure 6.2 to successfully transmit even voice messages.

Figure 6.3 is the classic wave-shaped carrier. Figure 6.2 is a carrier generated by a chaotic process. We could get into the interesting field of chaos cryptography to see how it works, but suffice to say it is a long story that

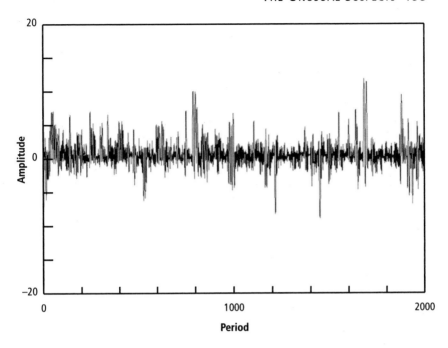

Figure 6.2. Detecting a signal from a mystery statistic.

involves things like phase-locking of chaotic signals. (We'll not dwell on that.) The point is that something that looks like noise can carry an orderly signal. Indeed, it can hide the orderly signal.

The key idea is that Figure 6.2 isn't really the result of a noise process at all. It is entirely determined as the solution of a system of nonlinear differential equations. That is true of climate too, in principle. It has governing equations that we cannot solve in a satisfactory way, as was outlined in Chapter 3. It is the sensitivity to initial guesses that make forecasts impossible for any long time, and it makes small things, below the resolution of climate models, have potentially big effects.

The best treatment for chaos was also discussed in Chapter 3. It was to work with averages, but as we have repeatedly stressed, not all averages solve this problem and particular strategies such as T-Rex raise other issues that in fact *add* to the problem. They are only our best chance for getting at a solution we may never get, but at this point it is all we have. After all, the torturously chaotic solutions of the Navier-Stokes equations are from equations that are themselves already averaged. So averages are in no way

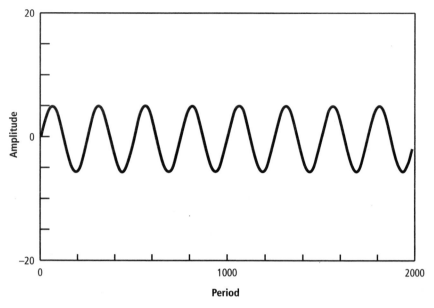

Figure 6.3. Classic shape of a signal carrier.

guaranteed to get us off the hook because it is quite possible to average over chaotic systems and end up with averages that are themselves chaotic.

The equations that produce Figure 6.2 are much simpler than the Navier-Stokes equations. We can plot the solution (in its phase space) to reveal what is known as an *attractor* for the chaotic system of equations. This attractor was named "the poodle." Well, it does look like the poodle was run over by a truck, but Lorenz's famous butterfly attractor didn't look like a butterfly either. That had two wings; this has four feet, a tail and a head. Also it is woolly like a poodle. It came from work that Weiguang Yao, Chris and Pei Yu were doing to find ways to defeat attempts to crack chaos encryption using deterministic methods.

The power of the poodle is that, unlike other chaotic attractors, it is woolly. Figure 6.5 shows the back of the poodle in closeup, so you can see it as it really is: not just a solid black mass, but something filled with strands of "hair."

These strands are actual solution paths, or trajectories, which in projection appear to cross each other. The poodle was quite resistant to cracking because of this feature. More particularly, it would be resistant to cracking by any sort of hammer blows from classic statistical methods.

Figure 6.4. The poodle attractor produces Figure 6.1.

Statistics can help if you are trying to detect a signal on a simple carrier like Figure 6.3. But on a carrier like Figure 6.5, they are no help, primarily because there is no way to separate the signal from the carrier solely on the basis of statistical properties. You need the exact equations to correctly produce the solution to unravel the signal.

In climate, we know we have a chaotic carrier, but we do not know what it is, because we do not have the equations for climate. That is, we do not have a theory for climate. If we are going to try to detect some signal it is carrying, we are out of luck. We do not know what the carrier would do if we were not here.

Furthermore, if one were to apply statistical methods, the classical assumptions that are routinely used do not generally apply to turbulent flow. In fact they do not even apply to the time series in Figure 6.2. We will demonstrate this in Chapter 7. There we will discover that Figure 6.2 is not likely a stationary random process, even though there is no trend. Stationarity, which we discussed in Chapter 5 briefly, need not involve apparent trends.

Finally, it is not enough to say that we can at least detect the presence

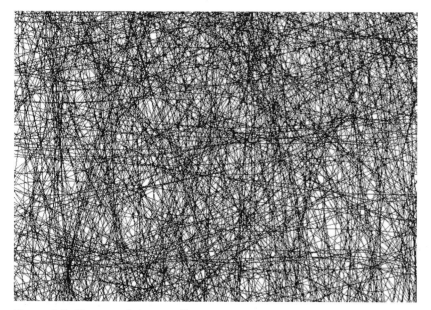

Figure 6.5. Closeup of the poodle.

of the signal without actually discerning it. How do you know you detected the existence of anything, if the basic series does not follow normal statistics, but rather a custom statistics induced by the dynamical system? That too will be considered in Chapter 7.

That is what PUN has done. They have in essence claimed they could use their hammer on the poodle. But if they try, they will only get bit.

Effect without a Cause

We have talked about meteorological "oscillations." It was claimed that the Pacific Ocean could be in one state and then suddenly switch to another while remaining steady before and after the switch. This was also discussed within the context of the tropospheric temperature record.

It is a widely held prejudice that large, sudden changes have to be caused by something external. The human eye will look at a complex thing and interpret structure even where there is none. It will seek messages in those structures that cannot be taken. Dynamics may be misinterpreted as an effect from an external cause.

Can "jumps" take place without something external causing them? Yes. In fact they arise in the simplest examples.

We discussed maps in Chapter 5. The logistic map is a famous non-linear map that exhibits chaos. It is very simple. Pick a number between zero and one (sounds like a card trick). Subtract it from one. Then multiply this by your original number. Take that result and multiply it by a special, fixed given number, called a coefficient, for your map. The result is the first computed value of your logistic map. You mapped the number you picked onto the result you computed.

People study where the map will take you. If you use your result and apply the same mapping procedure to it, you will get a new number as a result. You can apply the mapping procedure to that number too. You can do it over and over, each time applying the mapping procedure to the last number.

What you get is a sequence of numbers, each one being the result of applying the map to the number before it in the sequence. This sequence is to a map what the solution is to a differential equation. It can be thought of as a model of many things just like the differential equation's solution.

Now let's create a new map from the basic logistic map. Suppose you take your first number and apply the logistic map three times. This application of the logistic map three times is a map too. So you can create new maps from the old ones. In this case, the rule is to apply the old one three times. Of course this seems like a trivial difference, because it just picks out every third number in the sequence produced by the old map. But it is still a legitimate map. We will call this the three-times logistic map.

The three-times logistic map has a sequence associated with it too. If your map coefficient is slightly less than one plus two times the square root of two, then you will get a plot like Figure 6.6 for this sequence.

Suppose Figure 6.6 represents some condition of the world from year to year. Everything is fine and uninteresting until you get past 350 years and then suddenly something happens. A jump takes place. Everyone is thinking that something must have caused this. They start to look for causes exterior to the world. But you know better. There is no external cause, even though people will offer many candidates for one. The jump is entirely a property of the internal dynamics. In this case, it is problematic to tell exactly when a jump will take place, but you know that one will.

In connection with the real climate, we mentioned the Pacific climate "oscillation" or shift above. We instinctively associate discrete changes to human interventions in nature. It is all too easy in that kind of exercise to end up hunting for human causality. Figure 6.6 shows a pretty simple

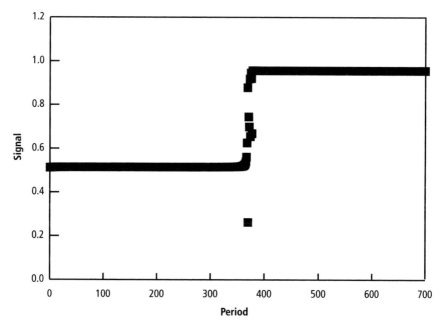

Figure 6.6. A "causeless" jump from the three-times logistic map.

chaotic process that for its first 350 values could lead observers to rhapsodize about the "delicate balance" of nature. But in a nonlinear world, big events happen naturally too.

There is a well-known average temperature that looks like this—the record from weather balloons. From 1958 (when the series starts) the average is flat up to about 1977, then jumps to a new level, where it sails along flat to the present. Again, the Pacific Climate Shift has been suggested as a cause of the jump, which may just be a label to put on the chaotic behaviour of the atmosphere-ocean system.

The climate is composed of many more complicated, interlocking nonlinear systems. Chaos is a pervasive feature of all of them: In fact it was through the study of such systems that chaotic behaviour was first found.

If the simple logistic equation shows such propensity to radical change in response to no external perturbations at all, why should we expect otherwise of the climate system? And why look only for "large" causes, or indeed for any external cause? The climate can vary, jump, rearrange itself and generally defy all expectations on any time and spatial scale. There is no requirement to "explain" it by an external mechanism. The internal dynamics of the climate system are more than adequate to account for just about any climate change, no matter how large or sudden.

At this point, the Big Panel may chime in by saying: "Precisely! The climate is a chaotic system, so changing something like the CO_2 concentration may produce drastic new behaviours and prompt disastrous climate alterations." But using this reasoning, you do not need large-scale emissions of infrared-absorbing gases to reorganize the climate system. A dog scratching his left ear might be held to account, or nothing at all. When a system shows a sensitivity to small perturbations, and these perturbations can rearrange the entire system, "small" and "large" lose their meaning in terms of simple measures. These things only mean something in terms of the mechanisms that make up the processes. If carbon dioxide emissions can provoke a climate change, so can nothing at all.

Maybe the climate regularly goes through changes in regime. They do not have to be sudden. Maybe they take place over many time and space scales. Maybe the climate "breathes," to use the language from above.

We looked at "big" suspects like the sun and the moon, because one instinctively looks for big causes of big effects. But we have also shown that little guys can cause big trouble too. And there are lots of little guys. In fact, there are an infinite number of them, and all of them can in principle cause changes on the scale of climate.

Not only do we have an unlimited number of suspects, we do not know if a crime has been committed at all. We have no way of knowing, even in principle, what the 20th century climate would have looked like if no one had ever learned how to extract and use fossil fuels. Yet much of the debate between "skeptics" and "believers" in global warming seems to be based on the assumption that we know what the climate would have looked like (in particular that T-Rex, however computed, would lie flat) and we just have to hash out whether this or that temperature statistic is really going up or not. If only it were that simple.

Signal Detectives and the Degenerates Who Would Not Confess

If we really are in the position of having so many sufficient causes of any observed change that we cannot even begin to attribute cause, how is it that the Big Panel can confidently claim things like "most of the warming observed over the past 50 years is attributable to human activities"? There are several types of argument used here. One works on the premise that data generated by a computer simulation of the climate are accurate *but for* a counterfactual change in a specific variable.

For instance, if a GCM produces data showing what some aspect of the climate would have been like had carbon dioxide emissions not risen after 1901, and the data thus computed are different from observations over the 20th century, then, according to the argument, we can attribute such differences to the change in the CO_2 level. Of course this requires the assumption that the model is an otherwise accurate representation of the climate system. If this premise is wrong, then one cannot conclude anything from the fact that the model-generated data differ from the real-world data.

An example of such an argument is Figure 14 in the Big Panel's Technical Summary. Another is a paper that appeared in *Science* in November 1998.[4] In that paper, the authors examined a pattern of what are called autocorrelation coefficients in a T-Rex series. These are parameters showing how averages at one point in time correlate to averages at successive points back in time. They found that a T-Rex from temperature stations has one autocorrelation pattern, but T-Rex generated by a climate model with reduced CO_2 levels shows a different autocorrelation pattern. *Obviously*, they inevitably conclude, this is evidence of a human influence on the climate:

> Assuming that the O/AGCM [climate model] control-run data provide a reasonable representation of the unforced behaviour of the real climate system, then a marked difference between the observations and the control-run results would provide evidence of external forcing effects in the observed temperature record.

Of course an alternative explanation for discrepancies of this or any other kind between model predictions and observational data is that the model is just a model, not an Enchanted Computing Machine, and does not provide a reasonable representation of the unforced behaviour of the real climate system. That would be the view in normal science.

Two critics of this study (Tsonis and Elsner) wrote to *Science* pointing out that two unrelated dynamical systems can still have identical autocorrelation functions, so it would not have proven anything whatever the authors found. Another critic (Singer) looked at the study data and found that the contrast between autocorrelations in the model-generated data and the observed temperature data also appear in a comparison between pre-1935 and post-1935 observational data. Applying the authors' interpretation means that there was a "greenhouse" effect pre-1935 that vanished after 1935. *Science* agreed to publish the comment by Tsonis and Elsner, but refused to print the one by Singer.

Another type of evidence uses procedures that attribute changes in average temperature statistics to various "forcing" agents, such as infrared-absorptive gases and solar variations. Climate models are employed to calculate estimates of what they call natural variation, as well as what are called signals: the supposed effects on T-Rex associated with each forcing agent. The real signals are unobservable, of course, so models must estimate them all. The model-generated signals are then compared to T-Rex using linear regression.

Five signal detectives from the UK presented this procedure in a June 1999 paper in *Nature*. The paper makes a confident promise in the title: "Causes of twentieth-century temperature change near the Earth's surface." Not *some possible* causes of changes in *some* temperature averages, mind you.

For the investigation, they chose a small number of climate-forcing suspects (infrared gases, sulphate aerosols, solar variation and volcanic dust), which were lumped into various combinations, with a model-generated signal computed for each combination. Using linear regression, they calculated the best-fit values of some scaling parameters that weight each of the signals so that the sum of the scaled signals, plus the natural variability, which is also an output of their climate model, resembled T-Rex as closely as possible. The climaticide squad examined the scaling parameters to see if they were statistically different from one. If a signal's scaling parameter is statistically "close" to one but not "close" to zero, then its effect is said to be "detected."

Why did they only interrogate a couple of possible suspects? Signal detection studies could not, even in principle, simultaneously consider all the small-scale forcing agents or the free changes that require no forcing. Even when restricting attention to the big suspects, the list was arbitrarily short. No signal detection study that we have ever seen includes in their lineup all the big forcing agents we have told you about in this chapter. And of those that include the role of the sun, they typically use a single aggregate measure like a flux reconstruction. But that hardly captures all the interactions between it and our climate.

Again, signal detectives rely on climate models to generate the data that will be used to test the assumptions behind... climate models. Obviously, to produce the signals and the natural variability range, the model that supplies the natural variability and the forcing signals must already be programmed with a certain climate sensitivity to each of the forcing variables being studied. But these sensitivities are the very scaling parameters that the climate squad is using their detection methods to estimate.

Hence any evidence they find is circular. The scaling parameter estimates from the signal detection regression are informative only if the forcing coefficients in the climate model were accurate to begin with. But if that were true, we wouldn't need signal detection studies! We could just look at the parameters in the model. If we do not trust the model parameters, then we cannot assume the model-generated signals and natural variability portray the actual climate, so the scaling parameters that come out of the regression model only tell us how well some uninterpretable numbers line up against T-Rex.

If you find that sort of thing interesting, that's fine. Scientists may have all sorts of reasons for studying these results. But they are ultimately just bits of information about the behaviour of models, not physical truths about the actual climate.

Deeper problems arise in the signal detection regression analysis. The technique only permits a limited number of signals to be tested simultaneously. Once signal detectives start adding more realistic levels of detail to the analysis, they run into something called *multicollinearity*. This arises when there are too many suspects who all look the same, and the computer cannot tell which ones matter and which ones do not. To put it more technically, it occurs when two or more variables are highly correlated, so they do not each add unique explanatory power to a statistical model.

It is routine in climatological papers to refer to multicollinearity as a "problem" that needs to be "corrected." But it is not a problem, it is a feature of the data. If all the suspects in the Great Train Robbery have altered their appearance so they look alike, this is a problem. Multicollinearity, on the other hand, is information that needs to be interpreted, not "corrected." In particular, its presence is telling the researcher that the signals are not unique and that it is not possible to connect changes in the underlying variables to changes in T-Rex with these data.

When two or more signals are not unique, they are sometimes called "degenerate." Signal detectives don't like degenerates. So they use a rough interrogation technique called "stepwise regression" to beat a confession out of the degenerate data. This method uses a numerical rule to rank the signals, then looks at each signal one at a time to see how much of the crime it will confess to. Whatever cannot be explained by the first signal on the list is given to the next signal on the list to explain. Whatever still can't be explained is passed further down the list, and so on.

Stepwise interrogation always extracts a confession—if one is to be had at all—out of the first signals on the list. So the key step is ranking the signals. Suppose there is a list of five or six signals to be tested. Ignore for a moment the problem that this list is arbitrarily short. Multicollinearity can be caused if any two (or more) are highly correlated, so suppose that two signals are correlated (degenerate). The stepwise regression rule uses a numerical formula to rank the signals. It is just a math formula, not a theory of the climate. The formula is used in lieu of a theory to determine what causes climate change. In this way, stepwise regression is a statistical attempt to get something for nothing. The proper way to deal with degenerate signals is to decide on theoretical grounds which ones can be excluded from the model and delete them, one by one, until only non-degenerate signals remain. Of course if we knew on theoretical grounds which signals belong in the model and which ones don't, we wouldn't need signal detection studies in the first place. If we have no special reason to delete one or more signals, yet multicollinearity gives us inconclusive results, this is not a "problem," it is a "discovery." It tells us that the signal detectives could not identify the culprit.

Conclusion Forcing

The idea of "forcing" has been forced onto the system. This language suggests a kind of metaphysical idea about climate change that has a mechanics reminiscent of the physics of ancient Greece. The climate state only moves when it is "forced," and when the forcing stops, the change stops too. But climate change is always happening, and it needs no external causes to keep it going. Any claim that things have been shown unequivocally otherwise is not justified.

With a case like this, it is no wonder our private eye had a bad feeling about what the climate cops were doing. There isn't a jury in the land who wouldn't overturn the conviction that carbon dioxide is the culprit behind the global warming menace. You might have noticed during our investigation, however, that we keep coming up against the idea that the climate problem is not solved, and that when you get into the details, it looks like we really do not have a clue about it. In their optimistic moments, some people think the evidence linking fossil fuel use to climate change is so solid it could stand up in court. We will take up that intriguing idea in Chapter 10. In the next one, we look at the question of how we should characterize the state of knowledge on the climate problem.

---

[1] Jacobson, Mark. "Strong Radiative Heating Due to the Mixing State of Black Carbon in Atmospheric Aerosols." *Nature* 409, 2001, 695–97.

[2] Friis-Christensen, E. and K. Lassen. "Length of the Solar Cycle: An Indicator of Solar Activity Closely Associated with Climate." *Science* 254, 1991, 698–700.

[3] Keeling, C.D. and T.P. Whorf. "The 1,800-Year Oceanic Tidal Cycle: A Possible Cause of Rapid Climate Change." *Proceedings of the National Academy of Sciences USA* 97, 2000, 3814–19.

[4] Wigley, T.M.L., R. L. Smith and B.D. Santer. "Anthropogenic Influence on the Autocorrelation Structure of Hemispheric-Mean Temperatures." *Science* 282, 27 November 1998, 1676–79.

7 Uncertainty and Nescience

In the Land Beyond Uncertainty

You may have noticed a recurring theme in our book, that the level of scientific understanding of climate change is very low. As in *very very* low. We are certainly not the first to point out the pervasive uncertainty in all aspects of this issue. The term "uncertainty" is found in many places in Big Panel reports, for instance. Chapter 14.2.2. of their really big scientific report says we don't have a clue what the climate will do, if you understand the jargon they use.

But when you think about it, this is a clue to the problem. It is possible, and nowadays even habitual, to acknowledge deep uncertainty in all aspects of the climate problem, then to turn around and talk about it as if one had a high level of certainty anyway. This is what the PUN has been doing for the past decade. Just recently, on July 10, 2002, our Environment Minister sent a letter to the editor of the *National Post* using this spin.

He was responding to an earlier article that talked about the limitations of climate models. He began by acknowledging that, yes, there are some "important challenges before climate models can provide us with... 'accurate forecasts.'" *Nevertheless*, he says, scientists are 66% to 90% certain that the "world will warm" by an amount ranging from merely unprecedented to catastrophic.

This seems to be classic doublespeak. But let us be fair here. The problem is not entirely created by those who speak it. It is a consequence of a failing in the language. We do not have the words in common speech to deal with situations like this.

The word "uncertainty" is not adequate for the situation. It suggests that what we possess is a modified form of "certainty." We are *certain*, with the modification "un." Anyone with a smattering of statistical knowledge will picture a bell-shaped curve depicting a range of known outcomes, where attention is appropriately drawn to the middle of the range as a sort of central most-likely-outcome. Reports from panels of the United Nations treat the situation as simply a gradation of certainty. UNcertainty, if you will.

But when discussing the core scientific issues in climate, we are not dealing with an accumulation of uncertain knowledge. We are dealing with an absence of knowledge. It is not just a lack of knowledge about the numbers associated with particular things. It is a lack of knowledge about what things exist to be uncertain about!

We have this vague idea of something called climate. We think we have some idea about some kind of average weather. But weather is not really a specific thing. It is a category of natural phenomena. Do we mean an average over a category of phenomena in which different things are measured in different units that cannot be directly compared? Of course not. Weather is filled with physical quantities. Which ones do we average? Which of an infinity of averages do we use? If we can decide, do we average them alone or do we average some arcane combination of them? Which time and space scales do we use? The questions just pile up, as we saw in Chapter 4.

Weather also implies dynamics. Averaging quantities is one thing. How do you average the dynamics? We talked about the problems the kinetic theorists and the turbulence theorists had with this in Chapter 3. When you do the wrong averages, you don't just get a bad number, you get something that may not even give you numbers. The result can "blow up," or you can get something "true" that you cannot use at all.

These serious problems don't stop people from averaging everything in sight in any old way and calling it "climate," and saying climate is doing this and doing that. But these averages of the things are not climate. They are just statistics from something we really don't understand. We have no way to say this average is the right one to use and that one isn't.

This ignorance is not measured in numbers. Albert Einstein is reputed to have said something about this: Not everything that counts can be counted and not everything that can be counted counts. There are plenty of things in mathematics and physics that are not numbers. Uncertainty about them is not a question of "how much," or even "which one," but whether what you seek even exists or whether you actually know what you are looking for. Climate is like that. The climate changes with which people have been concerned fare no better.

There is a word for this kind of lack of knowledge. It is *nescience*, from the Latin "not to know." We are uncertain of the human population of the Earth, but we can put a number with some uncertainty attached to it. But to ask a question you are nescient about, ask something where the answer depends on lots of assumptions that you cannot confirm or refute.

For example, how many extraterrestrials live on Earth? The direct answer to the question is a number, but most people would not believe that anyone has ever seen extraterrestrials. There is a question of existence: whether they have ever been here, assuming they exist at all, or whether they are here now, assuming they ever have been.

Then there is a question of what an extraterrestrial actually is. Suppose the aliens only send human agents they have trained elsewhere. Do we count them or not? You can't put a numerical uncertainty on this because the uncertainty does not depend on some measurement noise. It is not even clear what is being measured.

Nescience is therefore not the same thing as uncertainty. For one thing, people feel a need to use qualifiers with uncertainty, i.e., the Big Panel is "a little" uncertain of this or "very" uncertain of that. But qualifiers are not necessary with *nescience*. If we are nescient about the effect of carbon dioxide on local temperatures, no information is added by saying we are "a little" or "very" nescient. Also, "uncertainty" suggests that more study can be counted on to help the situation. We might speak of "bands" of uncertainty that can be shrunk by getting more data or a faster computer. But often a situation of nescience is so intrinsic to the problem under study that more data and bigger computers will not resolve the problem.

In this chapter, we will explore the distinction between uncertainty and nescience in climate. It would be nice if we only had to deal with uncertainty, but there is a lot of nescience as well, and it is pervasive in some key topics, as we have actually already seen. Mere uncertainty is the least of our problems.

What follows is an example where we have to contend with mere uncertainty: estimating past atmospheric carbon dioxide concentrations. Then we will compare it with examples of nescience on two questions: whether adding carbon dioxide to the atmosphere causes local warming and whether purely statistical methods can detect causality in global warming. More data and experimentation may resolve the first issue. But the latter two questions reflect the absence of theory. More data, faster computers and bigger laboratories cannot, on their own, resolve them.

Mere Uncertainty: Past Carbon Dioxide Levels

In talking about the carbon dioxide cycle, as we will shortly, we encounter a bit of odd terminology that we mentioned briefly in Chapter 3. Rather than discussing CO_2 itself, people discuss "carbon." Carbon on its own is not the issue, carbon with two oxygen molecules attached is. However, for some calculations, it has become customary to use carbon on its own as the standard to add up things that are hard to measure any other way.

Eventually everything gets put into units of billion tonnes (gigatonnes) of carbon, but since things have been converted from numerous different starting points, the word "equivalent" is added. So the units are "Gigatonnes carbon equivalent," or GtC. It is just a way of reminding the reader how the number came about. There is a model hidden in it.

The people who do these calculations try to be as accurate as they can, but at every step where something is estimated or converted into something else, more error and uncertainty creep in. So we never have anything better than rough estimates. Perhaps a better translation of GtC would be "Gigatonnes carbon in some form, more or less."

There are about 750 GtC of CO_2 in the atmosphere. The stock of CO_2 fixed as carbon in land biota (plants, animals and soils) is about 2,000 GtC, in the oceans it is 40,000 GtC, and in fossil fuel reserves it is 5,000 to 10,000 GtC. CO_2 is constantly being exchanged between the surface and the atmosphere. Plant respiration and decomposition releases and withdraws about 60 GtC (plus or minus two) annually into the atmosphere. The ocean releases and withdraws about 90 GtC (plus or minus two). These are very large additions and withdrawals from the atmosphere in terms of what we are believed to contribute in fossil-fuel-based emissions, which is only about 6 GtC, or 4% of the total land and ocean emissions. Minor variations in natural release and withdrawal can swamp anything that we may have contributed.

Thus the whole global warming scenario is a key-turning mechanism, as discussed in Chapter 6. Most of the effects we have on the world are like that. Here we have an extremely small effect coming up the middle between huge inputs and outputs that are nearly balanced. As long as these huge natural sources and sinks stay strictly balanced, the small contributions from fossil fuels can tip the balance, like a feather placed on one of two identical bowling balls balanced on a scale. But the large sources and sinks have not typically been balanced.

Since 1958, the CO_2 concentration in the atmosphere has been measured at an observatory at Mauna Loa, Hawaii. The current concentration is about 370 parts per million. A commonly used method for estimating earlier atmospheric concentrations of CO_2 is to examine the contents of tiny bubbles of air trapped by snow as it forms the ice cap on places like Antarctica and Greenland. Drilling out shafts of ice allows reconstruction of the chemistry of the air over the distant past.

Figure 7.1 shows a survey of estimated past concentrations based on Figure 9 in the Big Panel Working Group I Technical Summary. Looking back 450 million years, CO_2 concentrations were about 4,500 to 6,000 parts per million. Then, from 450 million to 300 million years ago, the level fell steadily to approximately current levels (300–400 ppm) and remained there for 100 million years. CO_2 levels then rose again and

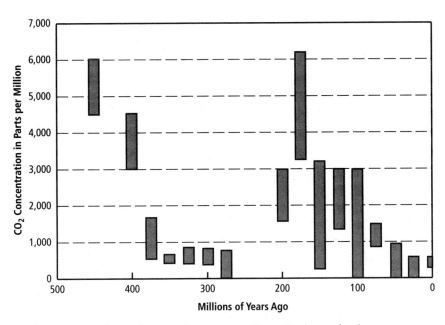

Figure 7.1. Estimated past CO_2 concentrations. Horizontal axis measures million years ago, vertical axis measures CO_2 concentrations in parts per million.

around 200 million years ago peaked at as much as 6,000 ppm again. The trend has been steadily downwards for the past 200 million years, with large periodic fluctuations. The lowest estimated CO_2 level during the past four ice age cycles (i.e., the past 420,000 years) was about 180 ppm.[1]

The big question is whether the current carbon dioxide level or rate of change is unusual compared to the past 50,000 years or so. On this point, the Big Panel maintains that CO_2 levels are rising faster and higher than anything seen for tens of thousands of years. But a look at the literature finds more uncertainty than this.

Numerous reconstructions of CO_2 levels over the past 40,000 years have been published since the early 1980s. The highest reported concentration is over 400 ppm, dated about AD 1700.[2] Subsequent surveys deleted this observation on the grounds that it is out of line with other estimates and therefore may have been contaminated or mismeasured.

In the same study, it was found that the estimated CO_2 concentration 10,000 years ago rose rapidly from about 200 ppm to over 300 ppm, with one sample showing over 400 ppm. Samples from two different sites (Camp Century and Byrd Camp) exhibit a range of 250 to nearly 500 ppm over the next 9,000 years.

The ice core from the Camp Century drilling had a concentration of about 420 ppm about 1,800 years ago, which fell to about 280 ppm at the 500 years ago point and rose to about 300 ppm over the next century. However, concerns that drilling fluid might contaminate the cores led these authors to recommend taking the lowest CO_2 values out of adjacent samples to represent the air at the time of the ice formation. Recent studies take pains to emphasize the so-called Taylor Dome series, which happens to be relatively smooth, showing a slow decrease between 10,000 and 8,000 years ago from 265 to 260 ppm, then a steady increase from 260 to 280 ppm up to pre-industrial times.

When an ice core is drilled out, someone must decide how to date each segment. If a shaft of ice was cut from a depth of 175 metres beneath the surface of an ice cap, it doesn't come with a date stamp: Someone has to decide not only how long ago the ice at that layer formed, but also how long ago the air got trapped in the ice. This second step turns out to be controversial. It takes many decades for bubbles in ice to close, during which time the surrounding air may flow in and out.

Hence an ice air bubble represents an average concentration over a very long span of time: as much as a century. This naturally smooths out any decadal-scale variability. It is conventional practice in studies of CO_2 in ice bubbles on short time scales—hundreds of years—to assume that air in the ice is 83 years *younger* than the ice in which it is trapped. (No, we're not making this up. See, for instance, the paper cited in endnote 2.)

It doesn't matter so much as you get to thousands and hundreds of thousands of years into the past what this correction is. But when we are looking into the past at a length of time on the same order as the correction, it is not so convincing. What if a different correction were used?

If the air were dated the *same* as the ice, recent samples would actually suggest that concentrations of CO_2 at the start of the 20th century were similar to levels measured at Mauna Loa in the 1980s. Figure 7.2 shows why. It's a graph of atmospheric carbon dioxide concentrations based on monthly measurements taken at Mauna Loa. Also shown is one measurement of carbon dioxide (denoted by *) from among the hundreds taken at the Siple Station in Antarctica.[3]

The asterisk on the left shows the carbon dioxide concentration of the air bubbles trapped in ice that is believed to have formed around 1900.

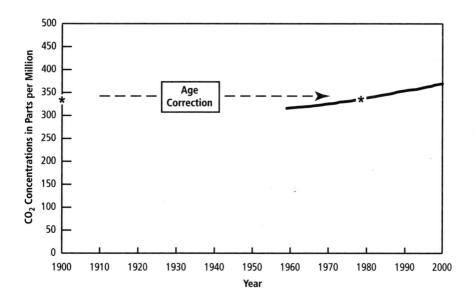

Figure 7.2. CO_2 concentrations measured at Mauna Loa observatory since 1959, showing ice-air age correction that places 1983 air in 1900 ice.

The air trapped in this older ice is presumed to be 83 years younger than the ice. Not 72 years or 23 years but 83 years.

If the air in the ice from 1900 were taken to be the CO_2 concentration in the air in 1900, it would mean that atmospheric CO_2 levels haven't gone up much. They would seem to fluctuate down and up by 25 ppm within a century time scale. But it is conventionally assumed that air circulating in 1983 was trapped into ice that froze around 1900, so this observation is shifted to the right and assigned to 1983 rather than 1900. Today's air is assumed to freeze into ice that froze around 1920.

While it is now conventional, this redating technique is certainly not universally accepted. One critic revealed a bit of nescience in the issue. Zbigniew Jaworowski is a physicist who published an essay[4] in 1996 pointing out that no experimental evidence has been produced to show how surface air can penetrate up to 100 metres into Antarctic ice, especially since in many regions the ice forms barriers of impenetrable crusts that separate the ice sheet into layers with no air circulation between them. Seems like a good point.

Physically, it would be unrealistic to expect that there would be just one time correction. Maybe this would not matter if you were looking at cores thousands of years old or even longer. But over time scales of a hundred years, a single correction must be regarded as artificial. Presumably there would be air from different times in various concentrations getting mixed together to make a kind of average, depending on the details of any particular piece of ice.

Another problem with the ice-core method is that it assumes the air drawn into the ice does not undergo any chemical reaction that would change the CO_2 concentration. The ice men published a study in 1999 discussing some of these issues.[5] The physics and chemistry of gas preservation at high pressures in ice caps is at present little understood. Dust and other impurities in ice can raise or deplete CO_2 levels in the ice as the layer develops. Ice cores taken from Greenland are now considered less reliable than those from Antarctica because they have much higher concentrations of impurities.

Within Antarctica, of the various locations of cores, those from the Taylor Dome are considered to have fewer artifacts due to impurities. There is no test that can conclusively show whether impurities have raised or depleted CO_2: Researchers just look for conspicuous variation across adjacent segments of the core. On these grounds, the ice-core samples from the Byrd Station are now considered less reliable because they show periods of dramatic short-term variation, while those from Taylor Dome, dated to the same time, appear very smooth.

Of course these arguments introduce a circularity in the CO_2 history reconstructions. We are told that the 20th century manifests unusual variability of CO_2 levels compared to past centuries. There are highly variable ice samples in the past too, but these are viewed as unreliable, partly because they are so variable. The 20th century is said to be unusual because none of the *reliable* ice samples show past CO_2 levels varying dramatically. And we can tell which ice samples are reliable: the ones that show smooth CO_2 levels.

The circularity extends to the ice-air age correction method: Nineteenth century air could not have had contemporary levels of CO_2 since the big increase did not happen until recently. Right? We know that the big increase only happened recently because 19th century air had a lower CO_2 content. Round and round we go.

As the trapped air is put under increasing pressure, the gases in it can *fractionate*, or separate into unmixed components. If these unmixed gases react differently with the compounds in the ice surrounding the bubble, the composition of the gas released when the core is drilled out will not be the same as in the air when the ice formed. The ice experts cited in endnote 5 claim that the Taylor Dome core is shallow enough (550 metres) that "this problem is not relevant."

However, Jaworowski pointed out in his essay that fractionation is not a simple function of depth, but is also influenced by the presence of liquid water and the salinity of water in the ice. CO_2 is highly soluble in liquid water, and increasingly so the greater the depth. At 180 metres, its solubility at $0°C$ is 14 times higher than at the ice cap surface. If there is any liquid water in contact with the gases in the bubble, this will deplete the CO_2 concentration, especially at greater depths, i.e., in the older samples. The ice bubble technique rests strongly on an assumption that there is no liquid water in the ice of Antarctica. Since there are known to be over 70 lakes under the ice cap, this is one of those many assumptions we just have to hope is true.

Clearly there are many possible pasts that would produce the data we have from the ice cores. Distinguishing among them requires making assumptions that only a time machine would render unnecessary. That is nescience, even though the assumptions themselves produce a number over time that is merely uncertain.

The Ice Men and the Leaf Lady

Examining air bubbles is not the only way to reconstruct past CO_2 concentrations. A more recent technique is to examine ancient tree leaves buried in peat. The density of stomata—small openings in the cell wall through which respiration takes place—varies inversely with the atmospheric CO_2 content.

In 1999, a team led by Friederike Wagner at Utrecht University published evidence gathered from collections of leaves recovered from peat bogs. They dated them to soon after the last ice age, from about 10,000 to 9,300 years ago. An advantage of studying leaf stomatal frequency is that it offers higher resolution in short time frames. That is, it can potentially pick up year-by-year variations that the ice core method misses. Wagner

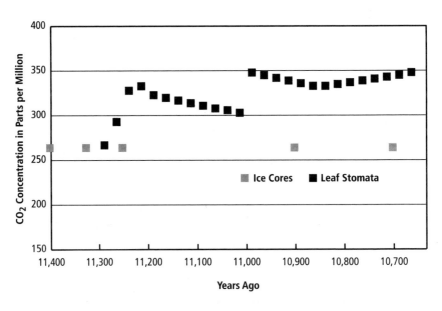

Figure 7.3. Comparison of estimated atmospheric CO_2 concentrations 11,400 years ago to 10,600 years ago. Blue circles show results from Taylor Dome ice core, green triangles show estimates derived from leaf stomatal index (LSI).

and her coauthors conclude that over the 700-year period they studied, atmospheric CO_2 rose from about 260 ppm to 327 ppm in under 100 years, then trended upward for another 500 years, rising and falling by tens of ppm over spans of as short as a decade, and peaked at almost 350 ppm around 9,380 years ago.

Wouldn't you know it, a different method gives different results. Scientists live for this sort of thing, because this is where interesting new discoveries are made.

The leaves challenge the view that atmospheric CO_2 was extremely stable at around 280 ppm since the end of the last ice age. Figure 6.3 compares the stomatal frequency-based reconstruction with the Taylor Dome ice core reconstructions. The trees tell a very different story than the ice: Not only was atmospheric CO_2 variable, it also reached contemporary levels shortly after the end of the last ice age.

This might lend support to the variable CO_2 history found in other Antarctic ice locations, such as the one originally proposed by Neftel back in 1982. Not surprisingly, the ice men disputed the findings of the leaf lady. An exchange between them was published in *Science* on December 3, 1999. While the teams clearly disagree, they could not explain the discrepancy between the two methods of reconstructing past CO_2 levels.

This is an example of uncertainty in climate science, but there is some nescience here too. There are built-in assumptions that cannot be confirmed or refuted at this point, so the errors in these numbers do not depend on observational noise, but on these nescient assumptions. The specialists in the field may never sort out the relative stability of CO_2 levels during the past 10,000 years within certain wide limits.

Nescience I: Carbon Dioxide and Surface Warming

The climate problem is not the first time that models have stood in for theory, only to turn out to be an unreliable substitute. Back in the 1970s, a lot of economists busied themselves building giant macroeconomic models of the economy. They had shown some promise in the early 1960s when the Kennedy administration tried basing policy choices on forecasts generated by computer models. But discrepancies between model predictions and observed events soon began piling up.

The models did not do well under the rising inflation of the late 1960s and early 1970s. When oil prices jumped in 1973, the models had difficulty predicting the effects. In the late 1970s, the phenomenon of "stagflation"—high inflation and economic recession—emerged. Yet the big models were based on a structure that assumed these two conditions could not coexist. When governments began running large structural deficits in the late 1970s and 1980s, the models predicted economic growth should occur, but instead the deficits appeared to be impeding it.

During this time, an economist at the University of Chicago named Robert Lucas published some essays exploring the role of expectations in economic decisionmaking. His point was that people do not just respond to *current* prices and conditions, but they look ahead and make plans based on their views of the future. The implications of this insight eventually turned macroeconomics on its head. It came to be understood that neither tax policy nor monetary policy could be relied upon to drive economic growth. As soon as people saw the government policy changes, they would change their savings and consumption decisions in ways that might easily cancel or even reverse the effects of the policy.

This revolution in thinking about macroeconomic modelling is today referred to simply as the Lucas Critique. The 60s-era macro models were based on empirical parameterizations of thousands of seemingly stable relationships in the economy: the proportion of income households put into savings, the amount of labour per acre used in agriculture, etc. The Lucas Critique pointed out that basic economic theory shows these parameters are dependent on, among other things, peoples' views of the current and future policy environment. If the government changes policy, the parameter values will change. Hence a parameterized model—no matter how big and complicated—is unreliable for evaluating the effects of a policy change, because the parameters are not invariant to changes in the policy regime.

One of the effects of this revolution in thinking about macroeconomics was the emergence of a deep skepticism among economists about large-scale models. About the same time as Lucas and his colleagues were exploring the problem of parameter invariance, another group at the University of Minnesota was using very small-time series econometric models to produce forecasts that were proving to be more accurate than those from big macroeconomic models. They were also developing theoretical explanations about why big models could not be more accurate than their own (in other words, why adding more equations and parameters would not yield greater accuracy).

These insights have taken a long time to work their way through the economics profession. There are still consulting firms that use large-scale macroeconomic models to tell the government what will happen if they change this or that policy. But this is more habit than anything. The problem of parameter invariance means that economic models, no matter how big and complicated, and no matter how fast they run on a fancy computer, cannot do more than generate speculative projections.

As we saw in Chapter 3, climatology has yet to absorb this lesson. Climate models embody parameterizations in lieu of basic theories too, because the physical theory is too difficult and, at the level of climate itself, the theory does not exist. Does this really matter for understanding the relationship between carbon dioxide and climate? Won't these little details just cancel out? The answer is quite simply *we do not know*.

The parameterizations in climate models are not dependent on people's expectations of the climate, as was the case in macroeconomic models. But they almost certainly depend on climate, even though it is typically presumed that they do not. In any case climate, like people's views, is difficult to pin down on human time and space scales. Climate, like weather, is a category of properties of the world. It does not lend itself to being reduced to some variable that can be input into a model to determine which of a family of parameterizations ought to be used.

As a simple example, we can revisit the question of whether carbon dioxide must actually cause warming at the surface, the way the ambient heat prejudice demands. (You can find more about it in a paper by Chris.[6])

We see because our eyes detect light in the visible range. But modern night vision equipment can "see" in the infrared. Infrared is a colour too red too see. If you could see the atmosphere in the infrared, it would not look black at night but would be luminous. The radiation is being emitted by the gases in the atmosphere, including water vapour, carbon dioxide and the other infrared gases. If you were to look down at the top of the atmosphere from above in the infrared, the presence of these gases would mean you would not be able to see all the way to the ground.

The intensity of radiation from these gases depends on temperature. This infrared light is not all at one temperature, but is the combination of infrared light from all the temperatures found at all levels in the atmosphere. If you were to imagine taking the same amount of energy from a single temperature rather than from a collection, we could find a particular temperature that gives the same outflow of energy. This is called the effective emission temperature.

As you move upwards into the atmosphere, the air gets cooler, as mountain climbers know. Assume that the temperature decreases by 6.5°C per kilometre from the surface. This is a well-known and widely used figure to represent the change of temperature with altitude. It is only used under about 15 kilometres from the surface, which is the troposphere. With this simple relationship between altitude and temperature, we can talk about either one interchangeably. You could picture a table with two columns, in which altitude is paired with a temperature from 15 km up all the way down to the surface. So there is an altitude that corresponds to the effective emission temperature. This is called the effective emission *altitude*. It is as if all of the emissions are coming from that altitude.

They really aren't coming from that altitude. The temperature is not really a single value. Nor, for that matter, is the rate at which temperature decreases with altitude a single value. The temperature really decreases with altitude at lots of rates, not just 6.5°C per km. The rate can vary from below 4°C to as high as nearly 10°C per km. But by making a few simplifying *modelling* assumptions, we seamlessly moved from a physical discussion into a modelling exercise.

That is how modelling always goes. There is no bump in the road that tells you that you have travelled from physics land to modelling land. It is like crossing from Iowa into Nebraska. There is a sign on the highway that you can miss if you aren't looking. You just have to know the difference between theory and modelling to keep track of where you are.

Sometimes people get confused about where they are. They lose track of whether they are doing modelling or physics. They think that someone else has done the physics to provide a number that the modeller can depend on. However, the person who provided the number may have a different idea. The number could be just a parameter that was useful for some specific use.

That is what happened with the value 6.5°C per km. It became part of a U.S. government publication called the U.S. Standard Atmosphere. Day by day, place by place, the real atmosphere doesn't conform to the "standard" atmosphere, but if you didn't know that, you might think that "standard" really meant something important instead of a kind of cartoon of what you might measure on a particular day outside your door.

So the model goes like this. If you put more CO_2 into the atmosphere, it absorbs more infrared light. Look at the atmosphere from space in the infrared. It is now more opaque. You cannot see as deeply into the atmosphere as you saw before, so the altitude where the emission was coming from must seem to be higher up in the atmosphere.

It is important that the effective emission temperature not change if everything else is fixed, as it has to be set to produce enough radiation emission to balance the sunlight's energy coming into the atmosphere. But the place where that unchanged temperature must now seem to come from is higher than before because you cannot see as deeply into the atmosphere. So the effective emission altitude is higher.

This point does not say anything about what the rest of the temperatures did to achieve this. It only requires that the effective emission altitude is higher than before. We have also not said what kind of movements take place to achieve this new effective emission altitude. All we know is that it is higher than before. However, if we know how much it has increased in altitude, we can use it to try to figure out what the rest of the temperatures have relaxed to as a result of the change.

With classical global-warming calculations, this increase in the altitude will correspond to about 300 m or so. Since the altitude is about 300 m higher and you have the temperature there, you can then project back down from there, at 6.5°C per km, to determine the temperature at the surface. It will be about 2°C higher than it was before. If you want more increase, just let the effective emission altitude increase even more.

This was used in some circles as a *proof* that increasing CO_2 into the atmosphere creates "global warming." It was fuel for the heat prejudice in people's minds that "the temperature" can only increase at the surface.

But there is no "the" temperature, as we have discussed at length. How can "it" be required to increase if "it" doesn't exist? Where are the slippery parts of the model that allows such a thing to seem like it makes sense?

It's right under our noses. Much to the surprise of many people, 6.5°C per km is not a constant of nature. It is actually part of a type of model of how the temperature in the atmosphere would behave under rather simplistic conditions. Moreover, if there were no convection in the atmosphere, there would be no linear decline in temperature with height at all.

With no convection, the temperature would have to drop very quickly near the surface, then switch to more gradual decline at higher altitude.

This is what the specialists call pure radiative equilibrium. Radiative equilibrium is what would happen to the Earth if the atmosphere were not a fluid and were made out of Lucite instead, as we mentioned in Chapter 6. In the model, however, the slack between the model temperature behaviour and what would happen in radiative equilibrium is presumed to be taken up by convection.

But in reality, convection means that in the real atmosphere, the temperature-per-altitude number is all over the place. It is not a constant and not like radiative equilibrium at all. Many have misunderstood this, thinking that 6.5 was observed or that it was calculated from theory. It is neither. It has been observed to range from less than 4°C to about 10°C per km depending on local conditions; "6.5" is just a parameter in a parameterization.

So back in the model, if the number itself can change as local conditions change, why would 6.5 remain fixed when the effective emission altitude increased? In the framework of the model, the increase constitutes a climate change. Why would 6.5 be the best single number to use in a new climate, within the context of the model? Changing it would simply mean that the convective transport within the model changes along with everything else in the model. It would mean, essentially, that more transport from the surface might be induced by more energy. In the model, it would be like turning up a "global air conditioner," so the additional CO_2 causes cooling rather than warming.

Of course those holding the heat prejudice don't like putting energy into any other form than heat. But it would be hard to justify any assumption that the dynamical vertical transport of energy would be unaffected by a climate change.

What does this mean? Once you allow the number 6.5 to change, all bets are off as to what happens to temperature at the surface. It would need only to change to 6.2 to eliminate any warming at the surface at all, which isn't large in the range of 4 to 10. In fact, if you let it go to 6.1, you would have cooling at the surface! That would upset the heat theorists, but there is absolutely nothing wrong with cooling at the surface within this model. In this simple model, you can also get cooling by allowing a small variation in the fraction of sunlight that is reflected before it gets into the system. And there are other arbitrarily fixed things that can be changed too, to get nearly any outcome at all.

So this is the way out. The "smoking-gun proof" is shown to have no predictive value at all. Once you allow the things that are held constant, for no physical reason, to change, you can get models to do nearly anything.

Here is another way to think about it. You have often heard that climate models "predict" warming at the surface. Some commentators acknowledge there is a wide range of results that come from models, but point out that at least they all agree on some amount of warming, as if this is informative. What this section has shown is that even this assurance can be overturned with a simple change in parameterization.

The models show surface warming from adding carbon dioxide to the atmosphere because of their programming. They could yield surface cooling with different programming, without violating any physical law. All that is required is to allow things to change in the model that do in fact change in the atmosphere.

This is not uncertainty, this is nescience. Error bars are of no use in relating the model surface "temperature," which is well defined within the model, to the temperature field of the atmosphere and oceans, which do not have a uniform surface temperature. The connection is just a caricature. How do you put error bars on Mickey Mouse?

Nescience II: Statistical Detection of Causality Between Carbon Dioxide and Climate, or Truth is Granger than Fiction

Within some PUN circles, there is an awareness of a problem that we talked about in the last chapter, that using models to test ideas about climate change is problematic. The tests only tell you something about the real climate if you assume the models are the same as the real climate, which we know is not true. Consequently, some researchers have been looking for an alternative that relies purely on statistical methods. As currently conceived, such a test would look at whether carbon dioxide and aerosols are driving changes in T-Rex, using only data derived from instrumental observations and not from model outputs.

You may have heard the old line that "correlation does not prove causation." This is a very popular thing to mention in classical applications of statistics. You don't hear it very much in the physical sciences because that sort of statistics is not really very important there. Those scientists are suspicious of such statistics.

Since statistical methods look for correlations, this would seem to rule out drawing conclusions about causality. But there is an argument used in some time-series-analysis circles that allows for inferences about causation. The technique is called Granger causality, named after the economist, Clive Granger, who first proposed the idea. It was originally used to study economic time series that seem to move together when it is not clear which one drives the other. For instance, interest rates and exchange rates often seem to move at the same time. So do interest-rate changes drive the exchange rate or vice versa?

The Granger causality technique poses the question as follows. If we know the value of one variable, does it allow us to improve our forecasts of the other variable? If so, we might conclude that some kind of causal relationship exists between them. However, in this case "improve" specifically means shrinking the forecast errors, on average. Actually it is not an ordinary average, it is the square root of the average of the squared forecast errors. Professor Thermos might interject that this is another case where the theory tells us not to use the mean.

With this change in the definition of causality, it is possible to envision a very general approach to detecting a human influence on climate. Recall that in Chapter 5, we looked at the use of statistical mappings to connect unlike things. Granger causality makes use of a statistical mapping between the thing we want to predict and the thing we believe has the causal influence.

With the lubricant of "randomness" to handle gaps between the fitted values and the observations, we can always compute the parameters of such a mapping. The particular technique is called vector autoregression, but the jargon is not important. What is important is that the procedure always "works." Unless some gross error has been made in specifying the statistical model, a computer will churn out parameters of the mapping without a moment's hesitation.

Models are not needed, just good old-fashioned data. Nor do we need to worry about the particular equations involved: The statistics can tell us which shape of curves fits best. It seems too good to be true—and, as we shall see, it is.

Since the mapping always yields parameters, we need some procedure to check whether they are "large" enough to count. In the trade, this is called "significance." The computer will work out something called a "t-statistic," which is the ratio of the mapping coefficient to a measure of its dispersion called the standard deviation. If this ratio is greater than two (roughly), the parameter is said to be large enough to matter. Why two? It makes things wrong less than 5% of the time. Why is that good? Tradition.

This point packs in a lot of assumptions, the biggest of which is that there is some stable "true" value against which we are testing. But we are comparing some physical observations on things like carbon dioxide concentrations to T-Rex, which is not a physical variable. It is a statistic composed of a mix of indices that measure a non-physical thing, namely, temperatures averaged over times and places.

Whatever the results are cannot be interpreted in terms of the physical environment, since people do not experience T-Rex. We went over all this in Chapter 4. Because the definition of T-Rex changes over time and is not a coherent physical variable, there is no reason to believe a mapping coefficient ought to be constant over a time series sample. This makes testing for significance a problem.

As well, it is a feature of these computations that you cannot just pull out the particular thing you are interested in—say the atmospheric CO_2 level—and test it all by itself. You have to test it simultaneously with everything else that might be causing the climate to change. If you used the complete list of the "big" climate change drivers that we presented in the last chapter, you would have already included more candidates than anyone else has, but you would barely have begun.

A Granger causality test might very well detect a structural relationship between carbon dioxide and T-Rex. It might also detect a relationship between the number of white boxes (as shown in Figure 4.4) and T-Rex. But no one includes the latter measure in these sorts of models.

And beyond variables related to the big suspects and those related to the construction of T-Rex, how would you include a measure of chaotic transitions? How would you test for the effects of small things that have big effects? And it is not just small things *today* that might be influential, it could be small things 10 or 20 years ago that matter. You have to include the influence of those as well.

How would you decide what you can leave out? If you leave out something that belongs in the analysis, its absence will distort the computations about what *is* included in the analysis, unless what is left out happens to be uncorrelated with everything else that is included. This is a big if, and we have no way of knowing if it is true or not, unless you compute the correlations. But if you had the data to do that, you would not be worried about leaving them out of the analysis.

Furthermore, including everything that might potentially cause climate change would require more data than you could fit into any computer. And once you have included a few dozen candidate variables, you will find they do not have unique explanatory power, since there is just not enough going on in T-Rex to need explaining.

Even so, perhaps these are all questions of uncertainty. But there is a deeper problem at work that mere data cannot fix. It pertains to the question of "significance." We spoke about those t-ratios. They work by computing probabilities of hypothetical outcomes.

To compute a probability requires doing some statistical modelling. Computers are good for this because they can crunch numbers. But they will only do what they are programmed to do, and if we do not know how to program them to do the right calculation, then speed and size are no help.

Statistical modelling begins with the concept of a distribution function. Most people have seen the sort of curve you get if you graph the distribution of, say, grades on an exam. There is a cluster of people between 65% and 75%, then steadily fewer going up toward 100% and down toward 0%. Many phenomena seem to occur in those distributions. The curve is thick in the middle and thin on the ends. There are equations that can be written down that describe those bell-shaped curves. The equations are called density functions, probability-density functions or simply distributions.

A particular density function is often presumed for statistical modelling, which means it can be used to compute probabilities of events within the model. The most famous is called the Gaussian or normal distribution. It is also called the bell curve. It is what is used when people talk about "belling the marks" on an examination.

To plot a normal distribution, you need to choose two parameter values: the mean and the variance. The mean tells you where the curve sits on an axis and the variance tells you how dispersed or wide it is. Figure 7.4 is a graph of two curves. The shorter one is the normal (Gaussian, or bell) curve. We will talk about the taller one below.

Knowing the density function is the backbone of the classic statistical methodology. Everything comes from it. All the statistics (averages, variances, t-statistics, etc.) that people like to compute in the classic statistical approach come from manipulating the density function itself.

Among the infinity of possible density functions, the Gaussian is king. A joke among statisticians is that mathematicians think the Gaussian density is correct because it is proven by experiment, and the experimentalists think that it is correct because it is proven by a mathematical theorem.

The reason for the interest in the Gaussian is something that is known as the central limit theorem. This states that no matter what the density function is, as long as you can compute the mean and the variance, if you have enough data and you average it in the right way, the average will fit a Gaussian curve. There is some small print to this guarantee, i.e., "some conditions apply." But the theorem opens up such attractive possibilities that to many users these conditions hardly seem important.

In reality, there are many cases in the world where the conditions for the central limit theorem don't hold. In these cases, the Gaussian density is not correct, and it does not become more correct just by adding more data. And this makes everything that gets built on the Gaussian assumption, including those t-statistics, a potential illusion.

But the statistics themselves will not tell you when you are caught in an illusion, you have to know on theoretical grounds. The Gaussian syndrome is impervious to refutation because the classical methodology will always produce something that seems to be the property of the data. Without a theory to tell you otherwise, there is no way to break out of the illusion. There is little awareness that using a Gaussian density is a modelling assumption that the actual data may not really support. In reality, there are many different sorts of density functions in the real world that are not Gaussian, nor do their averages converge to a Gaussian shape.

For example, consider a time series arising from something called a random walk (also called a drunkard's walk). The movement from one instant to the next is dominated by randomness. If the series can range over an open, continuous space, the positions of many random walkers from the start would form a Gaussian density. But not all spaces in nature are open and continuous. Many have a **fractal** geometry, which is a mathematical object potentially with non-integer dimension. A random walk on a fractal has a density function that is itself a fractal and not a Gaussian.

Random walks are especially pertinent because numerous studies in recent years have found temperature series behave like random walks, depending on the time scale considered.

If instead of walks, you also allowed your walkers to fly, then you would not get a Gaussian at all. You would get a broader class of density function, of which the Gaussian is a singular limiting case. These flights are called Lévy flights, and the resulting class of distributions is called Lévy stable. They are stable in the sense that they stay the same under certain linear transformations.

There are many instances of Lévy densities in the natural world. The most relevant for this book is the shape of spectral lines of CO_2 in the lower atmosphere. This is known in molecular spectroscopy as a Lorentz shape or profile. However, it is known among statisticians as a Cauchy density and it too belongs to the Lévy family. We have plotted it in Figure 7.4. It is the taller of the two densities.

The characteristic of Lévy densities is that they are spikey in the center and they have fatter wings than the Gaussian. You can see this in Figure 7.4. You would never be able to match the Gaussian to the Lévy density by changing the variance of the Gaussian. If you tried to match the wings, the variance would have to be increased, but this would decrease the centre, which is already too small.

The fatter wings are important. They mean that unusual events are

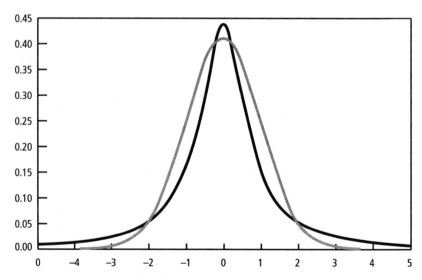

Figure 7.4. Two density functions.

more likely in the Lévy case. And it means another very important thing, not apparent just from looking at pictures or at the data. No member of the Lévy class, except the Gaussian density, has a finite variance. (Some members do not even have a mean.)

We said earlier that to apply the central limit theorem, you need, at the very least, to be able to compute the mean and the variance. While you can always take a list of numbers and apply the variance formula, this is just an estimate of how the density function is shaped. If you are sampling from a list of numbers that follow the Lévy distribution, you can make an estimate of the variance. But if you sample over longer times, your variance will be larger. As you sample over longer and longer intervals, it will keep growing with the interval. The Lévy process is not stationary.

Non-stationarity is typical of turbulent flow, which as you know is pervasive in the atmosphere. Lévy distributions also appear in astronomy. They have also turned up in models of rainfall statistics, among others. In other words, we have good reason to believe that in the study of climate, our Gaussian warranty is void.

In Figure 6.2, we showed a graph of data generated by a chaotic attractor. It is easy to imagine that it describes some aspect of the climate. What would happen if we applied classic statistical methods based on the Gaussian assumption?

We have plotted in Figure 7.5 a distribution of data of the same variable that would give the time series in Figure 6.2. It would be tempting to note the famous "bell" shape and try to model it as Gaussian. So we have overlaid a Gaussian curve, with the variance adjusted to match the peak of the density from the data. It is clear that the chaotic attractor from Chapter 6 cannot be a Gaussian process. The fat wings suggest a Lévy process, implying non-stationarity. If that were so, our conventional statistical methods based on the central limit theorem would generate illusory and false results.

The plot is certainly reminiscent of Figure 7.4. But closer inspection shows we are not really dealing with a traditional random process here. It is both chaotic and bounded. Gaussian and Lévy densities are not bounded, which means they have values (albeit very few) located even very large distances from the peak. But the "poodle" data of Figure 6.2 have no data located at great distances from the peak because poodles are bounded (see figure 6.3). There are poodles in many sizes, but none extend to infinity.

It is not Gaussian. It is Lévy-like, but it is not really a true Lévy

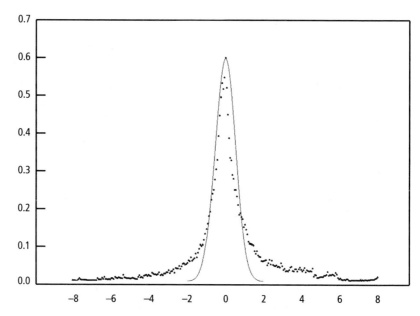

Figure 7.5. Densities from data of Figure 6.2 and the Gaussian density.

distribution because its fat Lévy-like wings suddenly stop. There is a fractal-like aspect to it too. But the point here is simply that the central limit theorem is not on, and anything you get from assuming it will be misleading. Perhaps some kind of principle to replace it might be possible for climate, but you need a specific physical theory for that, which we don't have.

When you treat the output of a deterministic system as random, nearly anything can happen in principle. If you are going to presume that it is Gaussian anyway, so that you can do some kind of "model-free" analysis, then you are just playing with terminology. The use of the Gaussian curve is a modelling assumption. Using it makes the statistical approach as much a modelling exercise as any of the computer models of Chapter 3 were. As a model, its ability to establish formal relationships, like "causality," for example, is just as questionable as any other climate models.

None of this would matter if what we were looking for was huge in comparison to the fluctuations. But it is the other way around. We are looking for a flea on the poodle, so these details matter enormously. If you want to have a scheme for finding change from normal behaviour, you have to know what normal behaviour is in the first place. About this we really have no idea. We are nescient.

A Concluding Comment

If uncertainty were our only problem, it would be hard enough to deal with, but at least we could hope that routine strategies—like getting more data and building bigger computers—would get us closer to the answers. But on the core issues our problem is nescience: the absence of understanding, not the presence of imperfect information. Bigger computers and more data will not guarantee anything here. The problems are of a different character.

Official Science is far away from recognizing this situation. In its "Action Plan 2000" (a climate-change-related policy blueprint) the Government of Canada's scientific advisors wrote: "Our scientific understanding of climate change is sound and leaves no doubt that it is essential to take action now to reduce emissions."

Whenever they are asked about the source of this confidence, their stock answer is that they are relying on the advice of the Big Panel. What must the Big Panel conceive its role to be if it could have cultivated this astounding mismatch between the real situation in the science and the beliefs of the government officials who rely upon it?

Talking about the limits of knowledge, and the difference between uncertainty and nescience, seems a fitting point to conclude our examination of the science side of climate change. We started talking, back in Chapter 2, about the human side of the climate change issue as it bears upon the enterprise of science. But then we turned our attention to the physical world, because it does not make much sense to talk about climate policy without first understanding the scientific situation. Now we find ourselves talking about the human side again. We hope you have gained an understanding of the physical issues sufficient to go beyond the simplistic certainties of the Doctrine. But they bring us to the big policy questions that haunt us still. In the remaining chapters of this book, we will turn our attention to these.

---

[1] See Petit, J.R., et al. "Climate and Atmospheric History of the Past 420,000 Years from the Vostok Ice Core, Antarctica." *Nature* 399, 1999, 429–36.

2   See Neftel, A., H. Oeschger, J. Schwander, B. Stauffer and R. Zumbrunn. "Ice Core Sample Measurements Give Atmospheric CO_2 Content During the Past 40,000 Years." *Nature* 295, 1982, 220–23.

3   See Friedli, H., H. Lötscher, H. Oeschger, U. Siegenthaler and B. Stauffer. "Ice Core Record of the $^{13}C/^{14}C$ Ratio of Atmospheric CO_2 in the Past Two Centuries." *Naure* 324, 1986, 237–38.

4   See Jaworowski, Zbigniew. "Reliability of Ice Core Records for Climate Prediction." In Emsley, J., ed. *The Global Warming Debate.* London: ESEF, 1995, 95–105.

5   Indermühle, A., T.F. Stocker, F. Joos, H. Fischer, H.J. Smith, M. Wahlen, B. Deck, D. Mastroianni, J. Tschumi, T. Blunier, R. Meyer and B. Stauffer. "Holocene Carbon-cycle Dynamics Based on CO_2 Trapped in Ice at Taylor Dome, Antarctica." *Nature* 398, 11 March 1999, 121–26.

6   Essex, Chris. "What Do Climate Models Teach Us about Global Warming?" *Pageoph* 135(1), 1991, 125–33.

8 Ceiling-Fan Gases and the Global Blowing Crisis

"Given the choice, I imagine nobody would opt for a world without any greenhouse, that is, a world with a mean temperature of about 259K [–14°C]. And probably few would opt for an ice-age world with a mean temperature of 275K to 280K [2–7°C]. To this point the greenhouse is seen as good. Further still, a majority clearly continues to see the greenhouse as good up to the present-day mean of about 290K. But, at the next 1.5K a drastic change of opinion sets in: The greenhouse suddenly becomes the sworn enemy of environmental groups, worldwide, to the extent that they rush off to Rio and elsewhere, and make a great deal of noise about it. I find it difficult to understand why. If I am told that computer calculations show immensely deleterious consequences would ensue, then I have a good laugh about it."
—Sir Fred Hoyle (1996)[1]

Waterworld and Other Impacts Studies

Items 5, 6 and 7 of the Doctrine tell us that Warming is bad, action is required immediately, and any action is better than none. From this action-packed theme has emerged a new genre of "impacts" studies. In Chapter 4, we pointed out the rise of the colourful usage of "impacts" in place of "effects" or "consequences" that has cartoonized scientific studies. Every possibility, no matter how banal, became couched in the language of collisions and blows. Like a cartoon, the reader could imagine the word-balloons of cartoons around every outcome: Kapow! Crash! Zowie!

Eventually speculative studies of significant consequences, with precious little scientific justification, came to be closer and closer to cautionary tales of dramatic fiction. The line between the two blurred to the point that movies came to be cadet versions of "impacts studies" in their own right. One of them was the 1996 film *Waterworld* with Kevin Costner.

As we discussed in Chapter 3, moviemaking can amount to modelling. In *Waterworld*, the modelling of global warming requires the polar ice caps to have melted. This so completely deluges the world there is no more land left. People are forced to live on dilapidated rafts tied together into floating cities, fending off pirates and looking for a mythical "dryland." This is impacts study at it its most dramatic, complete with cartoon-like action and sound effects.

Waterworld skeptics have pointed out that Arctic ice is already floating, so even if the poles melted, it would not raise sea levels enough to cover the continents. And the warming of the ocean's surface may cause an increase rather than a decrease in the formation of Antarctic ice—it is complicated! Moreover, if the earth's surface was covered in water, it could not absorb as much atmospheric infrared radiation, so local temperatures would start falling again. Most important, it cannot be sunny all the time in Waterworld, so it may be too cold for Jeanne Tripplehorne to run around in a thin T-shirt all the time.

Nevertheless, that "study" did raise public awareness and alert policy-makers to the importance of controlling infrared gas emissions to prevent an evil one-eyed warrior named Deacon from taking over the world.

That seems to be an important purpose of impacts studies: to alert policymakers. Following Costner et al, many more have been produced for this purpose. After the PUN released one of its SPAMs in 2001, the head of another UN agency declared it would "sound alarm bells in every national capital." There was no mention of whether it would summarize the science or instruct the public, but that might not have been the purpose.

While impacts studies all describe some aspect of the post-apocalyptic global warming wasteland, most do not attempt to characterize changes to the entire planet. Instead, they focus on specific regions, which requires that they take as their starting points the assumption that we have reliable climate forecasts at the local level. As we explained back in Chapter 3, we do not have such things now, and we may never have them. The theory is not there to support such computations. So there is no factual basis on which to proceed.

Nonetheless, this sequel to climate science has become quite popular.

Governments no longer just try to sell global warming policy on the basis of fears about T-Rex rising, but on the economic consequences of secondary phenomena that will supposedly take place. For instance, April 2002, Canada's Environment Minister, David Anderson, in went on a cross-country speaking tour to drum up support for the Kyoto Protocol. His discussion focused on the costs of these indirect effects, about which he was able to speak in impressive detail. In Calgary, for instance, he warned:

> Violent weather with more intense storms, damaged ecosystems, subsiding infrastructure, eroded coasts, more drought, more pestilence, more air pollution and increased damage to human health—these are just a few of the costs of inaction.

Notwithstanding his pretensions to detailed knowledge of the future, there is no scientific basis for believing these things any more than outcomes to the contrary. Like so many others, Mr. Anderson is simply caught up in the Doctrine of Certainty, playing his assigned role to give it voice and propagate it further. One way it gets spread is by funding impacts studies so more of this stuff can be produced.

Incidentally, in pro-Kyoto circles, the economic effects of climate change are nowadays called "the costs of inaction." The implication is that there is some kind of "action" open to us that would prevent all these perils. Since Kyoto would have no effect on climate change, as we will explain in Chapter 9, we think it should count as a form of "inaction" too. This is not what Kyoto supporters mean, but it is the logical implication of their choice of wording.

You might think it strange for a federal cabinet minister to travel across the country, frightening the public with highly specific predictions of impending social and economic disasters without having any solid information to back up such claims. After his speaking trip, it evidently occurred to the people at Environment Canada that it might not be a bad idea to begin putting together some information on the subject. So about two weeks after Mr. Anderson had finished his alarm-raising tour, Ross received the e-mail on page 236 from one of Mr. Anderson's staff.

It is clear from this message that the minister went cross-country on a speaking tour to explain to Canadians what his scientists say the "costs of inaction on climate change" will be, going into specific and dreadful detail

Date: 25/04/2002 4:45 p.m.

To: rmckitri@uoguelph.ca

Hello Dr. McKitrick,

I am currently putting together a file on the estimated economic costs of inaction with respect to climate change. I was given your name as someone who might be interested in the studies that will fall out of this file.

I have found an incredible lack of information on this topic in the Canadian context (and not significantly more in the American context). There is plenty of work being done regarding the costs of mitigation, but little on inaction.

Early next week, I will be meeting with some of our scientific researchers to determine what economic studies will be feasible in the short term and to determine a taxonomy to the studies that will be done in the future. We hope to get estimates of economic costs of climate change in the agricultural sector, transportation sector, and the primary resource sector to begin the list. Also, we hope to address particular problems such as extreme weather events, and sea level increases. This is by no means an exhaustive list.

Due to the scope of this project, we will be contracting out a number of studies. If you are interested in remaining in contact with me regarding potential future work or out of interest, please let me know. If you know of any other researchers currently looking into this field, or who have worked on these types of studies, please forward me their names or give them mine.

Thank you very much for your time.

Regards, [etc.]

at each stop, *before* his own scientists had begun assembling the information. Worse, when they finally do undertake the research, the minister will have already announced what the conclusions will be!

So the "scientific researchers" will just be asked to dig out the results that will back him up, which, by this point in the book, should be regarded as business as usual when science and politics meet. No doubt they will do so, and we can expect before year's end there will be a report from Environment Canada about the "costs of inaction on climate change," warning about the dreadful effects on agriculture, transportation, primary industries and more. It will make reference to problems arising from severe weather, rising seas, eroded coasts, pestilence and subsiding infrastructure. It will be official: frightful to the public, useful to politicians and appalling to regular scientists.

The Little Panel

The most important recent contribution to the impacts genre is the Third Assessment Report of PUN's Working Group II (Impacts, Adaptation and Vulnerability). Working Group II is a sort of junior Big Panel that takes the regional forecasts of the Big Panel and dreams up the problems that will be caused. To distinguish it from the Big Panel, we will call it the Little Panel.

The first challenge for the Little Panel is that the Big Panel does not produce forecasts, only "scenarios." Furthermore, since modellers have no confidence in their regional climate simulations (and say so), the Little Panel has to base its work on scenarios that their own creators doubt. That doesn't put the kibosh on anything, as it might if this were normal science. The Little Panel takes as given that global warming will be bad and has produced a long report on this theme. The claim is, in effect: We don't know what will happen; but it will be very bad.

They too produced a SPAM. It reads, in part:

> Projected climate changes during the 21st century have the potential to lead to future large-scale and possibly irreversible changes in Earth systems resulting in impacts at continental and global scales. These possibilities are very climate scenario-dependent and a full range of plausible scenarios has not yet been evaluated.
>
> (Little Panel SPAM 2001, page 6).

The paragraph then goes on to list some possible disasters, including the shutdown of the Gulf Stream, melting of the major ice caps on Greenland and Antarctica and a runaway "greenhouse" effect if melting Arctic permafrost releases vast stores of methane. But, they concede, the probability of any of these things is "probably very low."

Read the quoted paragraph again carefully. The second sentence is a sly way of saying "the outcomes we are studying are products of fictional scenarios, and we have been selective in our storytelling." As we will explain below, the climate scenarios are the product of "storyline" teams that have produced very pessimistic projections of the future, all of which overstate emission and fuel consumption trends over the 1990s. The "full range" of plausible scenarios ought to include beneficial outcomes, but these have been ignored. As Hollywood executives know, if you want a different storyline, get some new writers.

Extreme disasters from climate change are possible, as are great gains. But only the former outcomes are ever discussed—and they are presented repeatedly in lurid detail. Little Panel reports are tautologies built around and in support of the Doctrine. Global warming will be bad. How do we know? All the scenarios we studied show global warming will be bad. Where did the scenarios come from? We wrote them. Why did we only write bad scenarios? Because global warming will be bad.

We could play tit-for-tat by writing our own sunnier scenarios, but we have no interest in these games. Instead, this chapter just tells you what some people have found about the effects of climate changes in the recent past, what models can and can't teach us about the future and observational evidence about human adaptation.

In looking at the impacts of climate change, we continually find the picture is more complex and less frightening than the Doctrine makes out. In our view, the evidence suggests that, even if carbon dioxide emissions are causing some changes in the local climate that people experience, with an overall tendency for some regions to have warmer weather, neither the extra CO_2 nor the warming appear to be particularly bad things, once we take into account the ingenuity humans apply to their daily lives. To try to conclude anything in more detail would simply presuppose greater knowledge of the future than we or anyone else possesses.

Measuring the Future

Even if the dream of simply predicting the natural climate had not died (as we talked about back in Chapter 3), climate prediction would be impossible for anyone who works on the assumption that humans have a large climatic influence. The reason is that economic activity is also unpredictable: There is no social science that claims any particular success at long-term forecasting of social or commercial activity. Perhaps as a result, the PUN uses a semantic back door and only refers to its forecasts as "projections" based on "storylines."

But governments do not treat these as mere "storylines." They have been willing to expend huge political capital and to make expensive policy commitments in the belief that these projections are actually going to come true. This ignores the fact that numbers that quantify the future are made up, and other numbers that are functions of these made-up numbers are also, perforce, made up. In this section, we look at some of the details of these made-up numbers on which so much of the global warming scare has been built.

The simulations in the PUN's previous report, back in 1995, predicted a change in T-Rex of +1.5 to +3.5 Bleeps over the ensuing century. They assumed that CO_2 and CO_2-equivalent infrared gas concentrations would grow at 1% per annum for the next century.[2] Yet most infrared gas concentrations are stable or declining, except for CO_2, which is not increasing at anywhere near 1% per annum. Over the period 1960 to 2000, the average annual percent change was just under 0.4%. At no point in the available record has the concentration of CO_2 grown by 1% in a single year, let alone over an extended period.

Climate modellers do not design the scenarios to be simulated. Instead, they were developed in the Big Panel's *Special Report on Emission Scenarios*. It is widely called SRES, but we will just call it the Big Panel's Storybook. A team of "scenario writers" developed "storylines" that describe possible future world development paths. These storylines are then used to produce projections of economic activity and infrared gas emissions. As a matter of official policy, the Storybook team considers all to be equally probable and assigns no estimates of each one's likelihood. Figure 8.1 shows three of the CO_2 emission projections, the A1B and B1 marker scenarios and the A1FI scenario. These three encompass quite a range of possible outcomes.

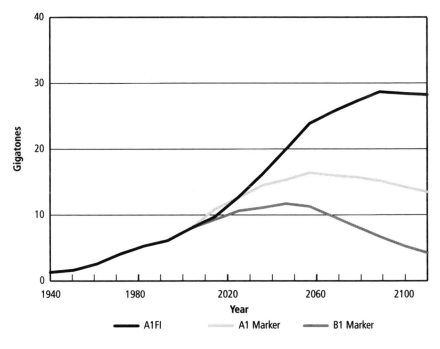

Figure 8.1. Past and projected future CO_2 emissions (gigatonnes carbon-equivalent) based on PUN scenarios.

Emissions are measured using so-called gigatonnes carbon equivalent, or GtC. Under the B1 scenario, total CO_2 emissions peak in 2040 at just under 50% above current emissions (11.7 GtC), while under the A1FI scenario, emissions more than quadruple over the next 80 years to over 28 GtC (compared to about six currently) and remain high thereafter. Emissions in 2100 under the A1FI scenario are over 6.5 times larger than under the B1 scenario.

The wideness of the range of global warming projections from climate models is to a large extent a result of the wideness of this range of emission scenarios. The A1FI scenario is the one that produces the extreme upper end projections, about +6 Bleeps in 100 years. It projects emissions of about 2,100 GtC over the next 100 years, which would be equivalent to the emissions from incinerating all living plants, animals and people.

As a guide to what might be a reasonable estimate of future emissions, note that for the past 30 years, annual global CO_2 emissions per person have been roughly constant at around 1.14 tonnes carbon equivalent.

With a world population of six billion souls, that gives us total emissions of about 6.8 GtC today. World population is expected to peak at about 10 billion in the middle of the 21st century, then decline thereafter.

Recent estimates suggest it may even peak sooner, at a lower total population, due mainly to declining birth rates around the world. But going with the conventional estimate, if per-capita emissions stay between 1.1 and 1.2 tonnes, we expect a peak of 11 to 12 GtC emissions in the middle of the century, declining thereafter as population falls. That would give us a scenario like the B1 storyline if we rule out technical change and energy efficiency improvements, which could lead to very different outcomes from that storyline. The A1FI scenario expects about eight billion people (as of 2080) to produce some 28 GtC annually, requiring a 2.5-fold increase in per-capita emissions. Burn baby burn!

One of the most important assumptions in these scenarios is the rate of growth of coal use. In an earlier group of scenarios, the IS92 family, global coal use was assumed to grow between 7% and 25% between 1990 and 1999. In the Storybook scenarios, coal use is assumed to grow over the same period by a minimum of 4% in the B1 scenario to a maximum of 31% in the A1FI scenario. Yet according to the data,[3] global coal consumption *fell* by over 10% between 1990 and 1999, so even the lowest scenario assumes 14% too much coal consumption. Despite this obvious and easily corrected overprediction of coal consumption during the past decade, no attempt was made to fix any emission scenarios prior to the release of the Third Assessment Report.

The projected increase in coal use for the three decades from 2000 to 2030 ranges from a low of 50% (B1) to a high of 160% (A1FI). By comparison, actual world coal consumption grew 40% in the three decades from 1970 to 1999, so the "low-end" scenario is still high compared to historical trends.

As for methane, the A1B and B1 scenarios project modest emission growth until somewhere between 2030 and 2050, after which emissions decline for the rest of the century without any policy intervention. The A1FI scenario predicts dramatic growth in methane emissions, to over double current levels by the end of the century. Methane emissions are very difficult to monitor, but atmospheric concentrations are observed in many places. The observed growth rate of average concentrations from 1990 to 2000 is below the lowest of the suite of storyline projections, it has been falling steadily for two decades and is now below zero (i.e., concentrations are declining). None of this has been reflected in the Big Panel's projections.

We looked in Chapter 7 at some of the challenges of estimating past carbon dioxide levels in the atmosphere. Estimating the future concentrations is even harder. Every year the oceans and the biosphere emit and absorb at least 20 times as much carbon dioxide as all fossil-fuel consumption combined (roughly 150 gigatonnes carbon-equivalent, compared to six or seven Gigatonnes from fuel burning). Minor changes in these natural processes, which we hardly understand at all, can have major effects over time on the composition of the atmosphere, and minor changes in the assumption of how long CO_2 remains in the atmosphere after release has a big effect on the predicted rate of buildup.

The A1FI File

But who cares about all these details, when the world is warming up faster than ever? At a widely publicized meeting in Shanghai in January 2001, the Big Panel leaders highlighted the astonishing result that the maximum projected increase in T-Rex for the next 100 years had gone from +3.5 Bleeps in their 1996 report to nearly +6.

This announcement drew headline coverage around the world—usually to the effect that the world was "warming much faster" than earlier expected. Commentators assumed that our acquisition of this new number resulted from an advance in our knowledge about the climate system.

As you can now imagine, that was not the case. The actual story is as follows. In the first draft of the Third Assessment Report, released in November 1999 for voluntary "peer review," four emission scenarios were included (A1, A2, B1 and B2). They yielded a projected change in T-Rex from +1.5 to +4.0 B over the next century—virtually unchanged from five years before.

We put "peer review" in inverted commas because it is not actually peer review the way most scientists understand the process. At a scientific journal, an article is submitted and sent anonymously to readers who pass judgement as to its suitability for publication. If the reviewers reject it, the editor will (usually) not publish it. If the reviewers request revisions, the authors are obliged to respond. Because the reviewers know their comments are decisive, they put some effort into the process.

In the Big Panel version of "peer review," chapters were posted on a password-protected Web site and a limited number of people were asked if they would provide comments. There were no guarantees that the reviewers' criticisms would be influential, especially if they had profound objections to the whole exercise. In the previous Big Panel report, there had been complaints from reviewers who had put many hours into reading drafts and submitting comments, only to find their submissions were ignored.

Another feature of ordinary peer review is the assumption that the document is not undergoing revision by the authors at the same time as it is being reviewed. Otherwise, the new or changed material gets inserted outside the purview of the reviewer. But while the PUN's scientific "peer review" was being done, another draft was being prepared and was released in April 2000, just before the end of the review period. The new material presented an envelope of 35 Storybook scenarios, including the A1FI scenario, for the first time, run through a simplified model "tuned" several different ways to mimic several general circulation models. These changes expanded the range of forecast increases in T-Rex to +1.3 to +5.0 B. A few weeks later, scientific "review" was closed. The new version was sent out for a final rewrite, which is done only by government bureaucrats.

The final draft was released in October 2000. To the surprise of many observers, the range of T-Rex projections had changed yet again, now spanning +1.4 to +5.8 B. The range was based on all the Storybook scenarios, but the projections had been redone using a simplified model tweaked with a wider range of core parameter values. The upper end of the projected change in T-Rex increased by nearly 2 Bleeps over the 11 months from November 1999 to October 2000, not because of any advance in climate science, but because of the inclusion of an extreme emission scenario and use of a wider set of parameter values in a simplified climate model.

One of the PUN's diligent peer reviewers is Dr. Vincent Gray, a climatologist in New Zealand. He followed the development of the upper-end projection, and in the winter of 2001 wrote a letter of protest to Dr. Martin Manning, vice-chairman of the Little Panel. Manning's e-mailed reply (February 9, 2001) was widely circulated among climate researchers and began as follows.

Dear Vince

    I do not have time to comment in any detail on your long message about scenarios and simple climate model projections. But very briefly...... [no text omitted here]

    You in fact confirm what I wrote in my report. The higher warming projections that arose toward the end of the TAR process are due to a high fossil-fuel emissions scenario rather than changes to climate models.

    Can I remind you that the fossil intensive scenario was not introduced by climate modellers or indeed anyone directly associated with the WG I report. It came from the SRES community and in particular was a response to final government review comments for the SRES. Many of us in the WG I community think the A1FI emissions are unrealistically high and if you read the atmospheric chemistry chapter you will see that they are projected to cause widespread impacts on air quality with direct health impacts.

    (. . .)

What is especially noteworthy is that the A1FI scenario was inserted on the orders of Official Science (government reviewers), not regular scientists, who consider them unrealistic. But you did not know this because you only ever hear from Official Science, and those folks have not felt it important to tell you. In the considerable publicity organized last winter by the PUN on the occasion of their new SPAM, much attention was focused on the high-end scenarios with the now famous +5.8 Bleep forecast. Throughout this process, the spokesmen of Official Science, including the climate modellers who allied themselves with the PUN to produce this report, kept silent about the origins of this number.

How Does Climate Matter?

The sort of temperature statistic that we graphed in Figure 4.1 seems to have risen by about half a Bleep over the 20th century. Whatever happened to the temperature field during this time, it was obviously not a catastrophe, since living standards and life expectancy improved considerably at the same time.

The distinction between T-Rex and the temperature field raises an interesting point. No one is talking about a change in the temperature field such that every place gets a little warmer all the time, even if that is the first interpretation that people have when they think of it. Over the past century, some locations in some very cold regions experienced a bit less cold at night in the dead of winter, while other areas like the tropics experienced little change.

Just looking at annual global averages creates the false impression that all regions warmed all the time, that daytime maximums rose along with nighttime minimums, etc. And this has created confusion at very high levels. Not long ago, the World Bank started publishing (in the *World Development Indicators,* no less) a set of national "carbon dioxide damages" estimates in which they take global emissions of CO_2, evaluate them at $20 per tonne of carbon-equivalent (for some reason), then distribute this sum across all countries according to population. They assume that everyone on the planet will be affected equally by climate change to the exact same degree. And that it will be bad. Now researchers, students and reporters around the world are at liberty to quote these numbers because, well, they're the *official* estimates, aren't they?

How does anyone know global warming would be bad for people? Here is one argument. Development economists have long noted that hot, humid countries tend to be poor, although for reasons that are not well understood. On that basis, some naive studies recently said that a rise in T-Rex of, say, +3 or +4 Bleeps will make them even poorer and will make temperate countries more like tropical countries (read: poorer). In this manner, a temperature statistic is presumed to drive the standard of living!

Where to begin? Well, for starters, even if we just take GCM simulations at face value, they say there will be relatively little change in the equatorial regions: The warming will mainly be in the polar regions. And warming there matters considerably less than elsewhere on the planet: If a spot in Antarctica is −30°C it will still be ice-bound at −25°C, and there will be very little in the way of a viable ecosystem (or economy) to be affected for better or worse.

Furthermore, it is not temperature on its own that impairs human productivity. It is more the daytime maximum temperature in combination with humidity. That is why meteorological agencies produce a "humidex" number, which combines temperature and the local relative humidity. If temperature rises but the air gets dryer, the net effect can actually be to make conditions more comfortable.

Of course, we only ever hear about temperature, and never about humidity. There is no physical justification for this. After all, enhanced infrared-energy absorption in the atmosphere could end up causing an increase in humidity rather than a rise in temperature. But while people swallowed the confusion about having a "global temperature," there seems to be little interest in tracking a "global" relative humidity. We never hear about a crisis of "global moistening" and the need to control emissions of "sauna gases," for which we can be thankful.

Looking at the opposite end of the weather scale, recall that in winter our meteorological agencies produce what are called "wind-chill factors." These combine temperature with air speed, because it is not just one or the other that determines how cold it feels. It can feel colder at –6°C when windy than –12°C when calm. So why don't we ever hear about the "global blowing" problem? After all, the energy absorbed in the atmosphere could manifest as kinetic energy (wind) rather than latent heat or increased temperature. So why doesn't anyone assemble a statistic of the "global wind speed," or invent catchy metaphors like "ceiling-fan gases"? Everyone would see right away that these are nonsensical abstractions and misleading metaphors.

So are, of course, "global temperatures" and the "greenhouse effect." Wind direction and velocity matter at a strictly local level, and GCMs cannot tell us how or if they will be affected by changing the infrared gas content of the atmosphere. Consequently GCMs cannot tell us how our experience of winter will change, even if we were to assume they could tell us something about average temperature changes. What if, for instance, winter air outside your front door is a half-degree warmer but a bit windier? The net effect may make you feel colder.

In any case, these days temperature and weather have very little economic effect. Most economic activity happens indoors, in climate-controlled settings. In advanced countries, outdoor production (mainly forestry and agriculture) accounts for less than three percent of national income. Manufacturing activities and financial and retail services make up the bulk of modern economic activity, and these are unaffected by outdoor temperature.

The Role of Adaptation

Changes in the ecosystem do not directly affect human welfare, but only in combination with the adaptive measures people take. The "cost" to society of winter, for instance, is the cost of adaptation and the residual cost of whatever inconvenience cannot be averted. But the averting measures relevant to any future climate change are very difficult to forecast since they may involve developing as yet unknown technologies.

Quite apart from the impossibility of forecasting what the climate will be like in a particular region in 100 years, the social costs of climate change are even more difficult to gauge, since we cannot predict the technologies for adapting to climate that will be available to society 100 years from now. When the Victorians attempted to forecast urban conditions after a century of growth, they worried about a coming crisis of manure accumulation in the streets and a critical shortage of space for stables. They did not foresee the development and use of the automobile. "Innovation" is inherently unpredictable, more or less by definition. But it is reasonable to predict that it will remain with us, given that it has always been with us.

Even if we assume technology will not advance, just taking account of currently feasible adaptation measures can dramatically affect the cost forecasts. Early simulations of the "cost" of climate change often overstated the case by assuming that people will be passive in the face of temperature and precipitation changes. This is not true in the short term, and there was no reason to believe it applies in the long term.

For instance, some early estimates suggested doubling atmospheric carbon dioxide concentrations could cut global cereal grain output by 20%. But these studies did not take account of the full range of adaptation measures people would employ. This is sometimes called the "dumb farmer" assumption. But one of the things economists have learned in recent years is that the way people adapt to changing conditions does not just reduce the costs of climate change, but can actually leave them better off than before!

Consider farmers who grow corn. They can choose among dozens of varieties of seed, each with different times to maturity. If you expect a long hot summer, pick a slow-maturing seed that does well in arid soil. If you expect a short wet summer, pick a faster-maturing and more water-intensive variety. Or hedge your bets by planting some of each or of different varieties altogether. And beyond corn there are dozens of agricultural and horticultural crops, each with dozens or hundreds of seed varieties covering every weather possibility. Should this not be enough to hedge your bets, farmers can buy crop insurance for weather-induced disasters. Now they can even buy "weather derivatives" on the Chicago Mercantile Exchange. These are contracts that promise compensation payments if weather conditions deteriorate to an agreed-upon level over a commercially relevant interval.

All these hedging options mean farmers can, for a price, at least maintain their base income regardless of the weather over a growing season. Changes in prevailing weather mean, at the simplest level, a small change in the optimal mix of crops and insurance instruments at each location.

But there is more to it. An optimal choice of seed variety for a new climate can produce improved results over optimal choices in the old. If the climate changes, you might lose if you don't adapt, but you may be better off than you were before if you do. Opportunities presented by a changing climate may be, on balance, harmful or beneficial. It is hopeless to pretend we can predict which will be the case a decade or a century from now.

The Ricardian Farmer

The analytical technique economists have applied to this topic is called *Ricardian analysis,* after David Ricardo, the great 19th century English political economist who, among other things, worked out a detailed theory of why rents on agricultural land take the values they do. Ricardo's insight was that if the market price for grain falls (say, by allowing more imports), the farmers who rent land will not be squeezed out of business. Instead, the landowners will end up cutting the rents they charge. The flip side is that banning grain imports does not benefit the tenant farmers but the landowners by boosting land rents. This argument was key to Britain's decision to allow free trade in grain in the early 1800s.

Today, Ricardian analysis relates past climate conditions to land values over time, and then, by speculating about future climate changes, we can forecast how the land values will change and thereby forecast net gains and losses for land users. The technique has been applied to all temperature-sensitive activities, such as farming, forestry, energy production, coastal activities and water utilities.

There are not many countries around the world that have adequate data to support such work, but the U.S. does. Three economists, Robert Mendelsohn, Bill Nordhaus and Daigee Shaw,[4] were the first to develop and apply the technique to U.S. agriculture. They found that, even ignoring the benefits of CO_2 fertilization, the sorts of climate changes projected by conventional models under global warming generate small net increases in agricultural land values.

Farmers will be more prosperous under the local conditions commonly forecast with global warming. Later, Mendelsohn and his coauthors extended the analysis to all sectors of the U.S. economy and again found small overall net gains for the U.S. under global warming.[5] They also extended Ricardian analysis to the global level (though with a much weaker set of land value data) and found a small net global benefit to the climate changes expected under a carbon-dioxide-doubling scenario.[6] Not all regions experienced a net benefit. Those that do include the northern countries like Canada, Russia and Eastern Europe, and parts of the U.S. The average global benefit is positive, however, suggesting that the winners could compensate the losers for their losses and still come out ahead.

In the case of forestry, early models predicted climate change would reduce U.S. wood production, but these did not allow for market participants to adapt to the changing nature of the ecosystem by adjusting harvest and planting activities. When these adaptive activities are built into dynamic models, the predicted costs vanish and productivity gains are forecast.

An issue here is the role of regional and international trade. No one is suggesting that changes to the climate would affect every place in the exact same way. If it happens at all, what it would likely do is rearrange favourable growing conditions for different crops. A century from now, some places that currently grow grain might be better suited to growing grapes, or vice versa.

What matters is that each region can take advantage of new potential climates, and to do so it must be able to make up through imports what it would no longer have a comparative advantage at producing, and earn revenue through exports on those items in which it would gain a comparative advantage. This also requires sensible agricultural policy that doesn't lock in the current cropping strategy through income programs that subsidize, say, oil seeds, and thereby make farmers less willing to switch into other crops as conditions might change.

Bugs and Bad Weather

The quotation early in this chapter from Mr. Anderson contained a word we don't hear much any more: *pestilence*. It sounds wonderfully gruesome and medieval. It really gets the imagination going.

> *Calgary, 2015: In the news today, dozens of hunchbacks in rags were seen pushing creaky wooden carts full of plague victims through the fetid ruins of our decaying city, crying, "Bring out your dead!" and kicking at the beggars and lepers who clung to them, pleading for alms. Scientists blame global warming.*

Mr. Anderson probably means that there will be more bugs, and with them the risk of more insect-borne diseases. The idea is that global warming would cause tropical diseases to spread into temperate zones. But it is already the case that insects travel across continents, sometimes through annual migration (e.g., butterflies) and sometimes through the slow expansion of habitat. Now that the West Nile virus is established in eastern North America, it is gradually expanding its domain westward and northward (to the dismay of Canadians!).

Cold weather is not what stops insect-borne diseases; public health programs stop them. In the 1920s, mosquito-borne malaria killed hundreds of thousands of Russians as far north as the Arctic Circle. As recently as 1950, malaria was experienced in large parts of the USA and Canada, and it was only eliminated from Holland in 1970.

Nor does hot weather automatically create the risk of insect-borne disease. From 1980 to 1999, 68 cases of dengue were reported in Texas. By contrast, in the three neighbouring Mexican states, 62,514 cases were reported over the same period.[7] The reduction of insect-borne diseases in developed countries is due to reduction in the extent of wetlands, application of insecticides, public health initiatives and other such programs. If we are concerned about global warming inducing a shift in the distribution of disease-carrying insects, direct disease-control methods are a more rational and cost-effective response than reductions in CO_2 emissions.

Honey, I Shrunk the Infrastructure

Then there was the warning about "subsiding infrastructure." Is this the same as "subsidized"? We have seen a lot of that kind of infrastructure lately. How does a highway overpass get "subsided"? One pictures it getting lower and lower until eventually it is an underpass. Perhaps the Confederation Bridge linking New Brunswick to Prince Edward Island will start sinking, to the point where motorists can only get halfway across the Northumberland Strait, at which point they have to transfer to a ferry for the rest of the trip and hope that the ferry doesn't subside before reaching Charlottetown. It is amazing what global warming can do.

Perhaps what was meant was that climate change will make the weather worse and that will damage our roads and bridges. We sure hope that roads and bridges are already being built to withstand rough weather. After all, storms and hail are not new. Perhaps what is new is the belief that they are under the direct control of economic planners.

There is no doubt that the more recent tropical storms in the U.S. have been the most costly ever in terms of damage done (the data are publicly available at www.nhc.noaa.gov/pastcost.html). Some folks, including governments and the Big Panel, have used this observation to suggest that weather itself is getting worse. However, the reason for the increased damage is not that storms are more frequent or more severe, but that there are far more people living in storm-prone areas, building more and more expensive homes and commercial enterprises.

This tells you something: People will voluntarily expose themselves to the risk of rough weather if, on other grounds, the activity that generates the risk is sufficiently beneficial. People move to subdivisions on the Florida coast because they like the scenery and they can earn a good income there. So even if—and we are not suggesting this is the case—using fossil fuels generated some bad weather trends, chances are, if we had the exact figures to do the cost-benefit comparison, we would conclude the tradeoff is worth it.

To use dollar measures of damage as indicators of storm damage requires adjustments for, at the very least, inflation, population and building density. One recent study that does so is by Roger Pielke Jr. and Christopher Landsea.[8] They cite many examples of media and government reports that claim hurricanes have been getting more frequent and more severe, based on superficial examination of damage-value data. But when the damage estimates are normalized for inflation, population change and fixed capital formation (buildings and equipment including housing), there is no clear trend over the century. The worst year was 1926, the 1940s and 1950s are the worst period overall, and there is a downward trend in normalized damages thereafter, except for a one-time jump in 1992 associated with Hurricane Andrew.

California or Bust

The U.S. is an interesting test case of people's preferences for climate, since it is likely the only country in the world in which people are free to choose among such a wide range of climates. When deciding where to live, a U.S. citizen can choose among tropical coasts, arid deserts, cool mountain ranges, mild coastal rainforests, hazard-prone floodplains, prairies with hot, humid summers and harsh winters, the northeast with its four distinct seasons, etc. Global warming is sometimes likened to obtaining a warmer, wetter climate with slightly more risk of extreme weather and drought. Is this the kind of climate people want?

Judging by the U.S. population movements in the recent past (see Figure 8.2), the answer is yes. From 1960 to 1996, U.S. population rose by about 47%. The Northeast and the Midwest added only 27% and 21% to their populations, respectively, over this time; in other words, they lost population share.

By contrast, the storm-prone South Atlantic coastal states grew by 114%, the mild and wet Pacific Northwest (Washington and Oregon) grew by 88% and California, despite its storms and mudslides, water shortages and earthquakes, grew by 102%—twice as fast as the nation as a whole. In proportionate terms, people chose not to live in the cool, predictable climates of the Midwest and Northeast, selecting instead the warmer, wetter regions, even with the risk of extreme weather and drought.[9]

The increase in voluntary exposure to riskier climates is possibly due to the fact that weather-related fatalities are not nearly as common as they once were, at least in industrialized countries. For instance, in the U.S., deaths due to hurricanes and tropical storms are negligible today com-

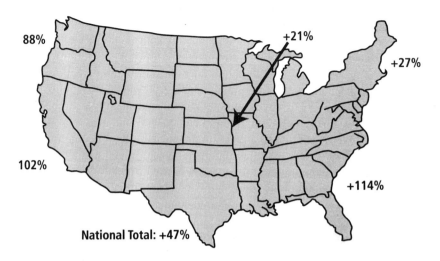

Figure 8.2. Population changes in the United States by region from 1960 to 1996 (according to U.S. census).

pared to the period up to the mid-1900s, despite the enormous increase in population located on the storm-prone southeastern U.S. coastline.

There have been 259 major Atlantic hurricanes recorded as making landfall in the U.S. over the past five centuries (as of this writing). The record of each of them is available from the U.S. National Hurricane Center at www.nhc.noaa.gov/pastdeadlya1.html. They are listed in order of deadliness. The first 39 entries on the list each caused more than 1,000 deaths. All the deadliest entries are for storms that occurred long ago. There are no entries for the period after 1981 until Hurricane Joan (October 1988), which is 95th most severe with 216 deaths. The first entry since 1995 is Hurricane Opal (October 1995), which is the 182nd most severe, with 59 deaths.

The dramatic decline in the deadliness of weather events in the U.S.— even with the increase in the number of people living in their path—can be attributed to better advanced warning, better building techniques and more resources available for adapting to the events. These are luxuries that wealthy countries can afford, and where they are absent, weather disasters continue to exact very high death tolls.

An important economic question, then, is whether it makes more sense to use resources reducing carbon dioxide emissions, in the vain hope that it will change the future of our climate, or to put resources into strategies to adapt to climate risks. It is no good advocating both. Dollars allocated to one end are unavailable for the other, so a decision must be made.

As for deadliness, in the U.S., at least, the number and severity of deadly storm events seems to be falling. Over the past century, there is no upward trend in the rate of tropical storms impinging on North America. From 1900 to 1959, there were, on average, 7.3 extreme hurricanes making landfall in the U.S. per decade. From 1960 to 1996, the average was 5.4 extreme hurricanes per decade (see www.nhc.noaa.gov/pastdec.html). In 2000, no hurricanes made landfall in the U.S. There is no upward trend in the frequency of Atlantic storms, nor is there any upward trend in the severity of storms. Moreover, the few GCM simulations that have attempted to parameterize severe storms under conditions of doubled CO_2 suggest that, if anything, their number will decrease and their average strength will not change.[10] Intuitively, this is consistent with the GCM prediction that temperatures rise more at the poles than at the equator, leading to shallower temperature gradients.

A recent study examined the formation of storms in the Atlantic over the 20th century, using hourly tide-gauge data from the U.S. coast.[11] These data allowed the authors to identify the number, duration and intensity of storm surges over the previous 100 years. They could find no discernible trend in storm activity and concluded that severe weather events are not responsive to the minor climate changes of the past 100 years. This is consistent with other studies from around the world, which find no trend increases in storm surges and wave height beyond that explained by natural variability.

Those Little Superscripts

All these points about weather and storms are made quite clear in the Big Panel's report. But this creates a puzzle for the alert reader of PUN documents. The Big Panel claims there is evidence of only minor changes in precipitation over the 20th century (pp. 142–43), no evidence of increased storm surges off the Atlantic ocean (p. 664), no evidence of increased storms and severe weather in the U.S. and elsewhere over the past century (pp. 162–63), nor is there any theoretical or modelling basis for predicting such increases in the future (pp. 573, 575). But the Little Panel made many such predictions when it released its summary in February 2001. The *Toronto Star* (February 18, 2001), in a story titled "Extreme Storms Darken Global Horizon," stated: "[The] Intergovernmental Panel on Climate Change will warn in a report to governments that the most serious threat to humanity isn't the forecast rise of several degrees in global average temperature by 2100 but the projected increase in the number and severity of floods, cyclones, droughts, heat waves and other examples of extreme weather." They then quoted a number of Little Panel contributors who played up this theme to dramatic effect.

What gives? To see how the PUN can send out such mixed messages, you need to learn to spot those little *superscripts*. On page 15 of the Big Panel SPAM, there is a table listing projected changes in extreme weather events over the 21st century. It seems to say that more intense precipitation events, droughts and cyclones are either "likely" or "very likely" over the coming decades. But read carefully. It actually says they are *likely*[7] or *very likely*[7]. Note the superscript. You won't find the meaning of "[7]" anywhere on page 15; it is back on page 2 of the summary. There you are told only that it refers to "judgemental estimates of confidence," which in the case of *likely* means a 60% to 90% chance and *very likely* means a 90% to 99% chance. This still doesn't say much (in fact it says nothing at all), and the "reader is referred to individual chapters for more details." Of course the "chapters" are in the big PUN report, which was not published at the same time as the summary—in fact, it wasn't released until about nine months later!

However, if you waited for the big report, and had the time and patience to look for details, you would eventually end up on pages 574–75. Here, in a nondescript portion of text, we are told what *very likely* means: "a number of models have been analyzed for such a change, all those analyzed show it in most regions, and it is physically plausible." Except in the case of fewer frost days, which are listed as "very likely" even though no models were ever analyzed for them. And *likely* means "theoretical studies and those models analyzed show such a change, but only a few current climate models are configured in such a way as to reasonably represent such a change." The only other category defined is "insufficient information to assess." But on page 575 it also says "climate models currently lack the spatial detail required to make confident projections" of extreme weather events like "thunderstorms, tornadoes, hail and lightning."

In fact none of them are simulated in climate models since they are all sub-grid scale phenomena.

And on page 573 it says: "We cannot reach any definitive conclusions concerning possible future increases in hail and lightning, and there is no information from AOGCMs [climate models] concerning future changes in tornado activity." The models that say something about changes in weather events are, apparently, different from the climate models that supposedly project global warming. The nature of these different models is left unexplained.

So there are three categories of future events: *likely, very likely* and *insufficient information to assess.* And now you know what *very likely*[7]

Table 8.1 Projected Changes and Events in Selected Areas over the Next 100 Years	
Phenomenon	**Probability of Occurrence**
A plumber named Mario will run an obstacle course through many cities, evading capture by strange creatures and other calamities.	*Very Likely*[9]
Waves of robot-like spacecraft firing laser weapons at random will invade the skies, but will be defeated by teenagers firing photon torpedoes from rapidly moving ground stations.	*Very Likely*[9]
Amateur drivers will attempt the Formula One and crash frequently, but drive away uninjured each time.	*Very Likely*[9]
The U.S. stock market will earn 50% per annum every year for the next century.	*Likely*[9]
The world will experience a devastating change in climate over the next century unless we put caulking around the windows.	*Likely*[9]

[9] The terms "Very Likely" and "Likely" are here used to denote expert assessment of subjective probabilistic magnitudes of integrated probability densities over rigorously defined sigma fields. "Very Likely" denotes the situation that all models which can potentially simulate such things do simulate such things in many cases, and numerous independent model runs were used to corroborate the events, even though the phenomenon itself may not have been proven theoretically definite. "Likely" means it is possible to configure a model of some kind that will generate this outcome, even though it is not bloody likely. The reader is referred to chapter 20 in *Chuckles the Clown and the Big Birthday Mystery* for more details.

means in the PUN report: Some computer models, of some sort, some-where, were able to simulate this in some places, and it is not physically impossible. Otherwise the event is merely *likely*[7].

In the spirit of fun, which we're sure the Big Panel intended when they pulled the *likely*[7] lark on everyone, we present Table 8.1. This lists some projected future changes in the world over the next 100 years. *Watch for those superscripts!*

If you would like to see other examples of how Working Group I has amused themselves with superscripts, check out the following: the mean-ing of the term "climate change"[1] (p. 2 of the SPAM), and the claim that 20th-century warming was unlikely[7] to be entirely natural in origin (p. 10 of the SPAM).

Après Global Warming, Le Déluge

A big fear associated with global warming is that sea levels will rise and inundate coastal areas. PUN scenarios of sea level rise range from 15 cm to 50 cm over the next 100 years. It is not the case that the central value is most likely and the extremes are progressively less likely: The probability density functions around these estimates are just not known. Since one of the more worrisome potential effects of climate change is the inundation of low-lying areas, the range of estimates corresponds to a large range of land areas potentially affected. A 15-cm change would require a relatively small investment in dykes and drainage machinery, while a 50-cm change could be much costlier, although in both cases the costs would be spread over at least a century.

Before presenting some material on sea level changes, we note the response of one sea level expert to the draft of the most recent PUN report on 20th century sea level changes (Working Group I TAR Chapter 11). The International Quaternary Research Society (INQUA) is an interna-tional academic organization devoted to the study of the earth's post-glacial history. In 1999, it convened a special commission on sea level changes and coastal erosion, headed by Professor Nils-Axel Mörner of the Department of Paleogeophysics and Geodynamics at Stockholm University. He was asked by the Big Panel to serve as an expert reviewer, and when he received a copy of the draft PUN report, Professor Mörner posted a series of delightfully frank comments on the INQUA Web site. They are reproduced in Figure 8.3. Remember that, as an official reviewer,

From the INQUA Web site (www.pog.su.se/sea/).

News & Views, No. 1, 1999/2000

IPCC has its Third Assessment Report (from its WG-I) out for external reviewing. I have just begun my reviewing. It is absolutely remarkable how inferior and one-sided this report is. And where are all the real sea level specialists from our Commission and from IGCP? They have had little or nothing to say in this report. If science is treated in this way, it is bound to go wrong. This being my first impression, I will finish my review and come back on the issue.
Informant: N.-A. Mörner, 1999

Chapter 11 on "Sea Level Changes" of UNIPCC's 1999 TAR paper (cf. above) was written by 33 persons; none of which represents actual sea level research. I have now finished a 7-page review report. It is a most shocking reading; lots of modeler wishes but very little hard facts based on real observational data by true sea level specialists. I allow myself a few quotations.

"It seems that the authors involved in this chapter were chosen not because of their deep knowledge in the subject, but rather because they should say what the climate model had predicted."

"This chapter has a low and unacceptable standard. It should be completely rewritten by a totally new group of authors chosen among the group of true sea level specialists."

"My concluding proposition is: (1) Dismiss the entire group of persons responsible for this chapter, (2) Form a new group based on real sea level specialists (e.g. INQUA), and (3) Let this group work independently of climate modeler."

Informant: N.-A. Mörner, 2000

Figure 8.3. Professor Mörner's note to colleagues about the Big Panel Study on sea levels.

Professor Mörner is one of the "top scientists" whose authority the Big Panel likes to invoke on its behalf merely because his name is listed in the back of the report.

There are long-term processes at work that affect sea levels. Sea levels have risen by over 100 metres in the past 18,000 years. Citing several recent studies, the Big Panel finds no evidence that the rate of increase has accelerated over the 20th century. Actually the point is made rather subtly. On page 663, section 11.3.2.2 begins with the statement, "Comparison of the rate of sea level rise over the last 100 years (1.0 to 2.0 mm/yr) with the geological rate over the last two millennia (0.1 to 0.2 mm/yr; Section 11.3.1) implies a comparatively recent acceleration in the rate of sea level rise."

Comparatively recent? What does that mean? It is not until you go down the page to the next paragraph that you read: "There is no evidence for any acceleration of sea level rise in data from the 20th century alone." This despite the fact that GCMs predict it should be happening. It turns out any "acceleration" of sea level rise was in the 19th century, but stopped during the 20th century, contrary to model predictions. Details, details.

Melting of Antarctic ice is an ongoing feature of the fact that we are not in an ice age anymore. Since the ice cap of the Arctic region is already floating, it cannot raise sea levels even if it does melt, as much of it does every summer. The Antarctic and Greenland ice sheets are landed, so they would affect sea levels if they lose mass, as they are expected to do over the next 6,000 years, until they are all melted or another ice age begins. But at present both areas are getting colder. And even if they were warming, any such melting will not happen for many centuries, with or without climate change.

One particular fear about sea level changes is that an enormous ice sheet in Western Antarctica, called (appropriately) the West Antarctic Ice Sheet, might collapse, releasing a great deal of ice and water into the world's oceans. Recent data[12] suggest that the four interglacial periods prior to the current one were warmer in Antarctica, but it is unlikely that the West Antarctica Ice Sheet collapsed during any of them.

On the other hand, the current interglacial phase is the longest by far, so the ice sheet may eventually disappear due to the *duration* of warming, rather than the level. That said, if the interglacial period weren't as long as it has been, there would not have been enough time for human civilization to emerge. Also, even in summer, the temperatures in Antarctica remain well below zero and the region is getting colder, so there is no apparent threat to the integrity of the ice sheet. But the only sure protection for Antarctic ice would be another ice age, and we are hardly rooting for that.

With respect to ocean-volume changes in a warming world, there are supposedly two opposing effects: thermal expansion and increased evaporation. What actually happens can't be treated theoretically because it includes all those inscrutable details of turbulence and thermodynamics we discussed back in Chapter 3. So in lieu of actually developing a theory of what goes on, there are some model parameterizations that work something like the following. As the oceans get warmer, the volume they occupy gets larger. But they also release more vapour into the air, some of which gets carried to the poles and deposited in the form of snow. These two effects may cancel each other out, or one may dominate. Until the fairy godmother gives us an Enchanted Computing Machine, we cannot run theoretical studies to determine which is the stronger effect. Meanwhile, empirical evidence is inconclusive. So all we have are artistic renditions in GCMs.

Nowadays these are drawn to show the Antarctic ice sheet as a whole expands in a warmer climate, rather than contracts. Warming would increase evaporation from the oceans, the moisture would be carried poleward and would accumulate on the ice cap as snow. This would occur faster than the ice melts at the edges. By contrast, the Greenland ice sheet is expected to lose mass due to warming. Some climate simulations suggest that changes in these two great ice sheets would largely offset each other.

However, a difficulty in constructing simulations is that the ice sheets are not in thermodynamic equilibrium with the atmosphere. That means there are ongoing changes that have nothing to do with any current climate change, but are driven by past gradients of temperature, pressure, etc. These natural dynamics must be characterized before accurate predictions of future changes in response to any infrared-gas-induced global warming can be made.

We saw one study[13] that presented a set of simulations that tried to account for background dynamics as well as future influence from climate change. They assume insanely high levels of carbon dioxide emissions and climate sensitivity, thus sketching a world crackling in a roaring red hell. What they call the "low" scenario assumes CO_2 concentrations grow at 1% per year, two and a half times the observed historical rate. In their model, this causes T-Rex to go up 2.4 Bleeps in 100 years. The "high" scenario assumes an eightfold increase in CO_2 concentrations over 100 years, which requires concentrations to grow at over 2% per year for the next century, five times faster than the observed average growth rate since 1960.

After controlling for background dynamics, these "illustrative scenarios" show sea level effects from Antarctica and Greenland that are still relatively small. Over the next century, says the scenario, sea levels are predicted to decline since the melting in Greenland will be more than offset by additional snowfall. The net melting of Greenland only begins about 90 years from now. Under their "low" scenario, they find the combined effect of Greenland melting and Antarctic expansion yields a sea level increase of about 6.2 cm after 140 years. Additional changes in sea levels might arise due to changes in the mass of mountain glaciers.

These are difficult to forecast, in part because there are 100,000 glaciers in the world and very few of them have been studied in enough detail to understand their response to temperature and precipitation changes. Over the next *one thousand* years, a modelled increase in sea level due to Antarctic and Greenland changes would be just over +55 cm. In each case, the Antarctic ice sheet expands and the Greenland ice sheet contracts.

Measuring a Moving Target

Mean sea level is rather tricky to measure. The North Atlantic Basin (covering much of Western Europe and northeastern North America) is undergoing a change caused by the retreat of the glaciers 10,500 years ago. The melting of these glaciers caused a heavy mass to be removed from the land. Like a sponge after a rock has been lifted off it, the land surface has gradually been rising: a process called post-glacial rebound.

However, the rise in northern Europe is causing southern Europe to sink slightly. So while Stockholm is getting higher, Venice is getting lower. For this reason, measures of sea level taken from northern hemisphere locations must be adjusted for the estimated effects of vertical land movement, which are large enough to play a role here. Of course this only matters for figuring out *absolute* sea level. From the point of view of those living along northern coasts, they are more interested in *relative* sea level: the height of the water relative to the land. If the land is rising as fast or faster than the oceans, there is no *practical* concern about sea level increase due to global warming.

From a measurement point of view, though, factoring land movement out of relative sea level changes is a very difficult geophysical problem. Absolute sea level estimates are the products of computer models that attempt to adjust for land shifts. As usual, there is a lot of uncertainty about how much to adjust the sea level data to account for land movements. The range of possible adjustment factors is about one third of the total estimated global increase in sea levels over the past 100 years.

In the southern oceans, good sea level data are available in areas where tidal swells are not too strong, land movement is minimal and large river discharges do not swell the coastal water level. The coast of Australia provides many good locations for measuring sea levels, and has the advantage of being near the region of the South Pacific where some small island states for whom concerns about inundation are acute are located.

It also has the advantage of being close to Tasmania, where long-time anti-Big Panel commentator John Daly lives. He compiled a complete set of sea level measures for the measuring stations around the Australian coast and posted them on the Internet.[14] Station records range in length from 23 to 90 years. Some stations show a slight increase in sea level, others show a net decrease, and the overall mean is less 0.3 millimetres per year, or three centimetres per century. This is smaller than the uncertainty about land displacement in the northern hemisphere. And considering all of the up and down in the oceans, among other details, it is unclear what the mean of water levels at places remote from each other actually means.

In any case, 3 cm/century is much smaller than the Big Panel's estimated 10-20 cm increase in global mean sea levels over the 20th century. The minimal increase in the mean sea level of the southern ocean, despite the alleged warming of the past century, raises the possibility that warming in the next century may not lead to as much of an increase as currently feared. It also raises a question about the accuracy of mean sea level measurements from the northern hemisphere. A fascinating discovery in 1985 has added to the mystery.

The Benchmark at Dead Man's Isle[15]

Off the east coast of Tasmania, near the town of Port Arthur, there is a tiny island called Dead Man's Isle (also called the Isle of the Dead locally). It was for a time a penal colony, but it was abandoned in the late 19th century.

In the 1830s, a local government official named Thomas Lempriere, who was interested in meteorology, began making systematic weather and marine records. It occurred to him that because of the local geography he was living in one of the best places in the world to record the mean sea level. In consultation with the explorer and scientist Sir James Clark Ross, he selected a site on the remote side of Dead Man's Isle and began a series of tide-gauge readings to determine the mean height of the water. In 1841, Lempriere and an assistant (and perhaps Ross himself) made a final determination of the mean sea level, then cut a surveyor's benchmark into the stone and erected a plaque indicating the date and time of the marking.

The benchmark was seen by a number of visitors over the next 50 years but was eventually forgotten. The plaque also disappeared. But in the mid-1980s, the mark was rediscovered by a local historian. At first no one knew what it was, but a search of island records yielded the answer: It is the oldest known mean sea level benchmark in the world.

What is remarkable, and what prompted the Australian government to launch a large and ongoing scientific inquiry into it, is that the benchmark sits about 35 centimetres *above* the current mean sea level. There is some dispute about whether the mark was cut at 2:44 p.m. or 4:44 p.m.—diary accounts of two visitors differ on what the accompanying plaque read—but this makes little difference, since written records found since then show it was sitting well above mean sea level by the end of the 19th century.

What the Lempriere benchmark apparently shows is that the surface of the southern oceans has fallen by about a foot since 1841, which contradicts the commonly held impression that they have risen by about 15 cm. The Australian Commonwealth Science and Industry Research Organization (CSIRO) was asked by the government to investigate the Ross-Lempriere benchmark. Their preferred scenario now is that the mark records the high-tide level rather than mean sea level. Against this view is the fact that the surveyors, and Sir James himself, described it at the time as a mark denoting mean sea level.

It is a fascinating tale. And here, as elsewhere, we find the sea levels topic is much more complicated (and interesting) than the Big Panel lets on. But there is a bottom line to which we find ourselves returning again and again when considering the costs of global warming: Humans are very ingenious at adapting.

Even if the sea level does rise, the worst-case forecasts are for something on the order of 50 cm over the length of time it takes for the concentration of CO_2 to double. At current rates of accumulation, that is about 170 years. It is quite conceivable that the world can build whatever coastal infrastructure is needed to deal with an 18-inch rise in the sea level, given 170 years to complete the task. After all, when the Red River in Manitoba flooded in

May of 1997, people there built a 25-km dyke at the town of Brunkild in just over three days. The dyke kept Winnipeg from being inundated.

However, building dykes is expensive and only wealthy countries can count on having the available resources to do it. So again the real issue is national wealth and economic resources: a point we shall return to several times yet.

Rising Carbon Dioxide Levels and Plant Growth

If you want to be tutored on the subject of the effects of rising CO_2 levels on plant life by biologists, we have four suggestions. First, there is a journal called *Global Change Biology* and pretty much any issue you pick up will get you started on this enormous topic. Second, there is a Web site run by Brookhaven National Laboratory (www.face.bnl.gov/findings.htm) in the U.S. that provides an online listing of research published on this topic. Third, there is an online bibliography at the Oak Ridge National Laboratory (http://cdiac.esd.ornl.gov/epubs/cdiac/cdiac129/cdiac129.html) on the subject of CO_2 and plants. Finally, there is a Web site (www.co2science.org) run by the Center for the Study of Carbon Dioxide and Global Change. It was set up by a U.S. Department of Agriculture biologist (Dr. Sherwood Idso) and his sons, Keith and Craig, each of whom has a Ph.D. in plant biology. They offer a well-organized library of information on studies showing how plants respond to CO_2 enrichment.

This latter one is not popular with Big Panel people because the Idsos have come to the conclusion that adding CO_2 to the atmosphere is actually a net benefit to the world. It is greening the planet, boosting agricultural and food productivity around the world, thickening forest cover, etc. And the plants that grow with extra CO_2 are more drought-resilient, less susceptible to stresses due to pollution and generally healthier. The Idsos make a point of stressing this in the editorials they include on their Web site, but they also tell you to where and how to look up the sources on which they base their conclusions so you don't have to take their word for anything.

So take your pick of information sources. They all give us the same general conclusion: CO_2 is good for plants. And what is good for plants is good for the rest of us, since we eat plants and we eat animals that eat plants. While the potential for CO_2 to be harmful, via changes in climate and weather, is distant and speculative and uncertain, the ecological benefits of enriching the CO_2 content of the air is close and well established.

There are of course people who will complain about this idea of CO_2 being helpful in any way. And there are those will pooh-pooh the research that plants improve because of it. But it is nice to know there is another side to such issues. Don't expect that in PUN reports.

Conclusions

This chapter asks whether the increase in carbon dioxide concentrations and other recent changes in climate are "bad" or "dangerous" things. We do not put much stock in the tautological approach of writing pessimistic scenarios, then concluding on that basis global warming will be bad. No one expects that "global warming," whatever it means, and should it actually happen, must be beneficial always and everywhere—just as no one could fairly argue that such a hypothetical thing is all bad. But the overall costs appear to be small and manageable, and may even be negative.

In the midst of all the present-day hype and hysteria, it is appropriate to ask hard but dull theoretical and empirical questions. What would small changes in climate actually do that would translate into effects that people experience in their everyday lives? How would people react and adapt? What happened in the past? What do we know about the effects of CO_2 enrichment of the atmosphere on plant life? Humans generally prefer warmer climates to colder ones and over the 20th century have sought them out, even at the risk of extreme weather events.

Meanwhile, storms and hurricanes are not increasing in frequency or severity. Relative sea levels do not appear to be rising due to atmospheric warming, and may even fall. The world's major ice sheets are not going to melt any time soon. And additional carbon dioxide in the atmosphere may well boost plant growth, which would benefit the whole food chain.

On balance, climatic change looks like it could be as much of a benefit as a cost, but the real issue is the extent to which societies continue to build economic strength. A policy like the Kyoto Protocol is intended to reduce economic growth in an attempt to slow down global warming. This is the wrong approach. Only wealthy societies can adapt advantageously to climate changes. A better approach to climate policy would be to continue pursuing economic growth around the world so that present and future generations will have the means to adapt and flourish in whatever climate they find themselves over the next century.

For some unfortunate reason, the term "adaptation" became a dirty word in PUN circles. Previous reports of the Little Panel did not pay attention to it. The most recent Little Panel report took a tentative look at it, but there is still a lot of cultural resistance in those circles. The problem, apparently, is that if we seriously look at the adaptation question and realize that it will be dead easy, the whole heroic enterprise of trying to reduce fossil fuel use and carbon dioxide emissions looks pointless, or even harmful. Nor would people be made to feel appropriate contrition for their sins if they could simply adapt to the consequences without much notice.

We find ourselves speaking of making people feel contrite, and trying to justify pointless but heroic gestures. What better segue to a chapter on the Kyoto Protocol?

---

1   Hoyle, Fred. "The Great Greenhouse Controversy." In Emsley, J., ed. *The Global Warming Debate*. London: ESEF, 1996, 179–90.

2   See the IPCC Data Distribution Centre, http://ipcc-ddc.cru.uea. ac.uk/cru_data/cru_index.html, (scenario guidelines). See also the UK Meteorlogy Office simulation page at www.meto.govt.uk/sec5/CR_div/ Anim/sul.html.

3   See www.eia.doe.gov/emeu/iea/wec.html for data.

4   Mendelsohn, Robert, William Nordhaus and Daigee Shaw. "The Impact of Global Warming on Agriculture: A Ricardian Analysis." *American Economic Review* 84(4), 1994, 753–71.

5   Mendelsohn, Robert and J. E. Neumann, eds. *The Impact of Climate Change on the United States Economy*. Cambridge: Cambridge University Press, 1999.

6   Mendelsohn, Robert, Wendy Morrison, Michael E. Schlesinger and Natalia Andronova. "Country-Specific Market Impacts of Climate Change." *Climatic Change* 45, 2000, 553–69.

7   Reiter, Paul. "Biting Back." Interview in *New Scientist*, September 23, 2000. Online at www.newscientist.com/opinion/opinion_225716. html.

8   Pielke Jr., Roger and Christopher W. Landsea (1998). "Normalized Hurricane Damages in the United States: 1925–95. *Weather and Forecasting* 13, 1998, 621–31.

9   All data are from the U.S. census Web site.

10 Bengtsson, L., M. Botzet and M. Esch. "Will Greenhouse-Gas-
    Induced Warming over the Next 50 Years Lead to Higher Frequency
    and Greater Intensity of Hurricanes?" *Tellus* 48A, 1996, 57–73;
    Landsea, Christopher W. "Comments on 'Will Greenhouse-Gas
    Induced Warming over the Next 50 Years Lead to Higher Frequency
    and Greater Intensity of Hurricanes?' *Tellus* 49A, 1997, 622–23.

11 Zhang, K., B.C. Douglas and S.P. Leatherman. "Twentieth-Century
    Storm Activity along the U.S. East Coast." *Journal of Climate* 13,
    2000, 1748–61.

12 Petit, J.R., et al. "Climate and Atmospheric History of the Past
    420,000 Years from the Vostok Ice Core, Antarctica." *Nature* 399,
    1999, 429–36.

13 Huybrechts, Phillippe, and Jan de Wolde. "The Dynamic Response of
    the Greenland and Antarctic Ice Sheets to Multiple-Century Climatic
    Warming." *Journal of Climate* 12, 1999, 2169–88.

14 See www.greeningearthsociety.org/Articles/2000/sea.htm. He
    obtained the raw data from the Australian National Tidal Facility at
    www.ntf.flinders.edu.au.

15 The story of Dead Man's Isle and the Lempriere benchmark is from
    John Daly's essay (www.john-daly.com/ross1841.htm).

9 Kyoto, Son of Doctrine

Flat Beer

Item 7 of the Doctrine of Certainty (Any action is better than none) implies that measures must be taken in response to the putative global climate change, and the nature of these measures is really unimportant as long as there are some. Anyone who has followed the discussion in recent years about global warming policy will have heard this recurring theme: *We must act now*. To consider whether the measures are worth doing, despite being ineffective, risky or detrimental, is considered a political position, rather than simple common sense. The risks and costs of acting are not considered, while the risks and costs of not acting are considered unthinkable. Thus the basis for making rational policy is driven out by the Doctrine.

If we are certain of how the atmosphere and oceans work and how they are going wrong, as the Doctrine says we are, then policy might be nothing more than implementing a global engineering solution. Indeed such solutions have been proposed. "Geoengineering"[1] implies ways of fixing the atmosphere and oceans, using the same simplistic metaphorical thinking that is the basis of the Doctrine: Solar radiation is heat, greenhouses work by the greenhouse effect, climate is one particular average temperature, etc. Within the assumptions of that picture, keeping some particular average temperature from rising seems only a matter of keeping unwanted solar radiation energy out, or trapping less of it at the surface. What could be simpler?

A number of ways have actually been proposed. They involve reflecting more solar radiation from the earth in some rather frightening ways, or limiting the radiation absorption due to carbon dioxide by altering the carbon dioxide budget. The former includes such bizarre and disturbing ideas as using artillery to fire enormous amounts of dust into the stratosphere; painting the roofs of the United States white; inserted tens of thousands of 10-km-wide mirrors in the earth's orbit; inserting billions of aluminized, hydrogen-filled balloons into the stratosphere; de-tuning jet engines of commercial aircraft to produce much more soot; burning sulphur on ships and in power plants to form sulphate aerosols to stimulate additional low marine clouds. One of the latter types of measures involves placing enormous quantities of iron into the oceans to stimulate generation of carbon-dioxide-absorbing phytoplankton.

Suppose these wild things actually can be used to control some particular global average temperature (and there is no guarantee they can do it effectively). It seems to have escaped many people that catastrophic environmental and climate changes can take place even without changing your favourite global average temperature. Tampering with something you don't understand like this is like trying to fix a computer with a hammer and good intentions. The potential for unintended nasty global consequences is obvious and ominous. Serious consideration of these dangerous scenarios is itself an unintended consequence of the relentless adherence to the Doctrine, in particular to the belief that we understand how climate works.

Surprisingly, the reason why these approaches are unappealing to many Doctrine supporters is not the clear risk they entail, but rather the notion that "fixing" things does not properly express suitable contrition for excesses in consumption. The planet, in their eyes, is not just being spoiled but despoiled, and those responsible should pay a price.

In this regard, it is curious that the Coca Cola Company and all other producers of carbonated beverages, from beer to champagne, have not become the target of political action. A bottle of soda pop has about two grams of carbon dioxide in it, and that amount will eventually be released into the atmosphere in one way or another. That can translate into several hundred thousand tonnes of carbon dioxide every year. Wouldn't it be consistent with the moral tone of "thinking locally and acting globally" to decline to drink carbonated beverages? Shouldn't these companies be required to produce flat beverages in the cause of stopping global climate change?

While the contribution of carbonated drinks to overall human carbon dioxide production is minuscule in comparison, it still amounts to maybe four million tonnes of CO_2 per year. Nonetheless, if the point is to achieve symbolic acts, and not to achieve a systematic balancing of costs and benefits, maybe flat drinks should play a role.

Fortunately, we need not fear the loss of our bubbly libations. Politicians and Official Science have come up with something that is similarly symbolic, and equally flat in terms of its internal logic, but which promises a far more costly penance: the Kyoto Protocol on Climate Change.

Diplomats from around the world, meeting in Kyoto, Japan, in 1997 to talk about global "warming," created this after two weeks of gruelling talks and marathon negotiating sessions. The good news is that in their sleep-deprived state the diplomats did not recommend geoengineering or controls on soft drinks. The bad news is that they embraced a strategy focused on changing infrared gas concentrations through small, outrageously expensive and ineffectual reductions in carbon dioxide emissions.

The Kyoto Protocol has had a troubled life since 1997. By the summer of 2001, it appeared almost certainly dead after the U.S. pulled out and Australia stated it would not go ahead without the Americans. But the parties met at Bonn, Germany, in July 2001 and, in a surprise turn of events, agreed to a slightly watered-down deal that at this point seems likely to be implemented. Canada's participation remains uncertain, and while Japan has ratified Kyoto, it has thus far ruled out most options for actually achieving its emission targets.

As this book has made clear, climate change is a very complicated topic. Unfortunately, with the Kyoto Protocol now on paper, many governments seem to believe they can sidestep the difficult task of thinking through the fundamental scientific and policy questions and instead focus on implementation. As we will see in what follows, in terms of an effect on climate, the Kyoto Protocol is actually little better than controlling the global consumption of soft drinks. If the climate is changing, regardless of the cause, it will do so to the same degree whether Kyoto goes into force or not.

This is one of the ways the Doctrine has already done damage. Time, energy and diplomatic goodwill that could have been invested in serious research on climate change and policy design have been wasted negotiating and analyzing this particular quixotic treaty.

This chapter looks at Kyoto in some detail. Even if, despite the previous chapters, the reader feels a continuing loyalty to the Doctrine, the Kyoto Protocol is not thereby justified on a rational basis. Any proposed policy must pass a cost-benefit test: The remedy must do enough good to justify the harm of implementing it.

Of course the Doctrine says that there is no cost for implementation. Environmental lobby groups have even come out lately and claimed that Canada could get rich by implementing Kyoto. This is simply not true. Regulation is an economic drag. By definition, it forces people to do what they would not have chosen to do, and there is always a cost for that. It may generate offsetting benefits, like the rule that forces everyone to drive on the right whether they want to or not, but sometimes the benefits do not exceed the costs.

If the regulations being proposed really could make us better off through some technology-driven boom, it would happen anyway, without the need for regulatory push. And while people are distracted by disputes over economics, it is widely missed that Kyoto will have a negligible effect on climate, whatever happens!

These considerations lead us to conclude that the Kyoto Protocol should not be implemented. In any normal area of policy analysis, this conclusion would be considered rational and uncontroversial. However, when the U.S. administration came to this conclusion and acted upon it, they discovered the hard way that, as with so much else in the field of global warming, mere rationality is itself controversial.

The Kyoto Protocol and its Gaseous Targets

Under the Kyoto Protocol, countries are grouped into "Annex I" nations and "Non-Annex I" nations. Annex I refers to the Framework Convention on Climate Change, the Really Big Treaty of which Kyoto is a mere sub-treaty, in particular a protocol authorized under Article 17 of the framework convention. (Follow all that?) Annex I is a list of nations, including the developed nations, who are members of the Organization for Economic Co-operation and Development, or OECD, and the members of the former Soviet Union (the latter referred to as EITs—economies in transition). They agreed to limits on emissions referenced to a 1990 base level. The U.S., for instance, agreed (at the time) that by the end of the implementation period, 2008–12, it will have reduced net emissions to 93% of 1990 levels. (Emissions of what exactly? We will return to that later.) Altogether, the Annex I nations agreed to a collective target of 5.2% below 1990 levels.

Non-Annex I nations made no commitments to future emission reductions! While there are undoubtedly sociopolitical motives for this distinction, in doesn't make much sense in terms of the overall emissions objectives behind the treaty. It's like fixing a leaking bucket by only blocking some of the holes. Actually it is worse. As we will explain below, plugging up one hole automatically makes other holes get bigger.

To become binding, 55 nations must ratify the protocol, including enough Annex I countries to account for at least 55% of Annex I carbon dioxide emissions. The signatories agreed to defer many of the most important details associated with implementation to future negotiations. For instance, the protocol makes some provision for trading emission credits among Annex I nations, but states that it must be "supplemental to domestic actions." An emissions-trading mechanism allows a country to exceed its allowable emissions level by buying permits from another country, which in turn must reduce its emissions by an equivalent amount.

An Annex I country cannot purchase credits from a non-Annex I country unless the latter first accepts emission reduction targets like an Annex I country. However there is a provision called the clean development mechanism whereby an Annex I country can claim credit for measures that help a non-Annex I country to reduce its emissions.

The protocol did not originally define how emission limits will be enforced, nor who would have the authority to approve trades. At their meeting in Bonn, participants did agree to a penalty system that punishes a country failing to keep its emissions within specified limits in one year by obligating it to reduce emissions an additional 30% later on. No financial penalties will be levied, however.

Russia and Eastern Europe are currently well below 1990 emission levels (owing to the economic collapse of the Soviet bloc), so these countries potentially have a lot of permits available for sale. Russia's block of permits are referred to as "hot air." Europe opposes allowing Russia to sell all these permits, and the agreement at Bonn included proposed caps on the amount of hot air that can be sold.

Another cap was imposed on so-called carbon sinks. This ill-defined term refers to the extent to which a nation's biomass sequesters carbon dioxide and any other infrared-absorbing gas. Sinks are extremely difficult to quantify, since doing so requires extensive information on land use changes over time with a lot of detail on forest age, tree species types and biomass, agricultural practices and so forth. The protocol originally only gave credit for new sinks which appeared after 1990. However, at the Bonn meeting, Canada, Japan and Australia obtained rather generous provisions for counting pre-existing sinks. (Even still, Australia later abandoned the treaty.)

Emissions are also difficult to quantify, especially because we aren't just talking about CO_2. There is no standard method for measuring net infrared gas emissions (other than carbon dioxide). Livestock, swamps, landfills, etc., for instance, produce methane, so precise monitoring is nearly impossible in much of the world. Another open question is how to weigh the relative effects of different infrared-absorbing gases. The protocol mandates the use of "global warming potentials" (GWP) to set collective effects of the different infrared gases. It does not identify appropriate formulas, which is not surprising since there aren't any.

In terms of climate change, the relative effects on climate of any particular infrared gas is virtually impossible to determine, as climate change has to have already taken place to even begin measuring. It is far from clear that this has even begun, let alone that it is already complete.

In terms of modelling, the lifetime GWP of a molecule of an infrared gas depends on the modelling assumptions in place, which pertain to every one of the vast array of imponderables of climate science, such as the turbulence problem. In terms of a global temperature statistic, which is not a genuine physical variable, it depends on what particular average of the temperature field is used, as discussed in the preceding chapters. For example, using balloon or satellite data instead of airport surface measurements would imply there is no effect for carbon dioxide. Similarly, modellers employ explicit and implicit definitions of T-Rex from their models.

So we have a curious situation. GWPs are impossible to quantify in a physically meaningful way, yet they are a fundamental policy target of a major international protocol. And countries propose to create billions of dollars worth of GWP-weighted financial instruments and trade them internationally. No wonder Enron lobbied so hard for Kyoto.

Lemon Permits

Let us overlook for the moment that the protocol covers emissions weighted by coefficients no one knows how to compute. The problems are only just beginning.

Students of international politics define a *stable* agreement as one in which all parties have a unilateral incentive to remain in the agreement, rather than defecting once it is in force. An example is a free-trade agreement among a group of nations. Since countries benefit from free trade, they have a selfish incentive to stick with it, even when aspects of it prove disagreeable.

This is why trade agreements have been relatively successful in the postwar era while other types of agreements have failed. For instance, international treaties for joint management of natural resources (especially fisheries) often fail because they are not self-enforcing. Consider a treaty to limit harvesting of a commercial fish in the open ocean. Each signatory is aware that "my" cheating on its own won't have all that big an effect on the entire stock, whereas if I comply and everyone else cheats, I'm worse off than if I never signed the agreement in the first place. Such agreements unravel, first through small-scale cheating, followed by angry accusations and recriminations and then open flaunting of the agreed-upon limits.

John Nash (of *A Beautiful Mind*) taught us how to think about these strategic traps. Consider the point of view of a country that achieves compliance with its Kyoto commitments. For Canada, that will require reducing carbon dioxide emissions by approximately 30% in 2010, at a projected cost of between 1% and 3% of annual national income. What exactly would be the benefits to Canada of this?

As we will discuss below, even if countries completely implement the agreement, there will be little change in the projected path of global warming. Eventually the public will figure out that there are no real benefits—even in principle—attached to the policies that are causing such economic dislocation, and ipso facto it is a net benefit to abandon them. Defection will follow—indeed it has already happened, and the treaty is not even in force yet.

The Leakage Effect

The so-called **leakages** problem is another obstacle. Suppose all OECD members undertake large reductions in fossil fuel use to control carbon dioxide emissions. This reduction in demand for fossil fuels will force down the world prices of fuels. Outside the OECD, non-abating countries will thus find it attractive to increase their use of cheaper fuel.

Also, energy-intensive industries will have an incentive to move themselves out of the OECD into the non-participating countries. Since energy use is relatively more coal-intensive outside the OECD, increases in energy demand will tend to be more carbon-dioxide-intensive.

The increase in non-OECD emissions as a percentage of the cut in OECD emissions is called the leakage rate. A leakage rate of 50% means that half of the emission reductions by participating countries are offset by newly increased emissions elsewhere.

The magnitude of these effects depends on economic parameters, such as the way the market responds to price changes and the degree of international capital mobility. One important study back in 1992 estimated a low leakage rate of under 4%, if there is no international capital mobility and if fuel prices remain nearly constant regardless of demand changes. When the model was changed to allow for capital mobility and price adjustments, the estimated leakage rate jumped to 50% for the first decade of implementation, then drifted down to a long-run level of around 26%. In other words, even ignoring problems of defection and cheating, between a quarter and a half of the emission reductions were offset by new emissions caused by the emission reduction policies themselves. This is what we meant by the picture of the leaky bucket, where fixing one hole causes other holes to expand.

Audits and Fibs

Yet another difficulty with the protocol is that countries are self-monitoring on both emissions and sinks. These are very difficult to measure even when done conscientiously, and Kyoto provides a not-so-subtle reward for nations that underestimate their net emissions.

For instance, suppose that country X reports net emissions at five million tonnes below that year's target and offers that many credits for sale on the permits market. Other countries may doubt country X's emission reports, but any audit must make use of figures provided mostly by country X itself. And if many nations are finding it difficult to meet their Kyoto commitments, they will be dependent on permit purchases, so they will have little incentive to audit country X's claims anyway.

One should not underestimate the incentives to cheat, even on international environmental agreements. The government of China has faced accusations that it greatly exaggerated recent carbon dioxide emission reductions. In January 2001, the PUN held a news conference in Beijing at which they, among other things, praised China for its spectacular progress in reducing emissions—14% in four years—supposedly proving to the U.S. how easily it could be done. Indeed it was easily done. In August, a World Bank-sponsored research project found Chinese emissions had not fallen much at all, and concluded the "official" numbers were bogus.[2] This happened even without a financial incentive for exaggerating emission reductions.

The Benefits of Implementation

Let us reiterate what is wrong with Kyoto. It is a treaty that covers quantities that cannot physically be defined or measured. It is inherently unstable and unenforceable. Economic leakages will offset up to half its effects even if everyone is honest. And if parties begin cheating, it is almost impossible to audit them, let alone force them to stop.

But Kyoto is a sort of diplomatic Rasputin that has been pronounced dead many times, only to stagger back onto the world stage, seemingly bigger and more indestructible than before. At least at the time of this writing, chances remain strong that Kyoto will be approved and implemented in some form by the end of 2002.

So we need to plumb its depths further. For the sake of argument, we will pretend that its many problems do not exist. What exactly could Kyoto accomplish?

Of course, to be worried about global warming you need at some level to believe climate model simulations. And they tell you that if warming is going to happen, Kyoto won't make a scrap of difference. The assumption one makes about post-2012 emissions affects the projections of the ultimate effects of Kyoto, though not by much. A well-known study by Tom Wigley[3] used the same models with which the Big Panel generated their warming scenarios to work out climate projections with and without implementation of the Kyoto Protocol. He considers three post-Kyoto scenarios, all of them rather optimistic.

In the first (called "NOMORE"), Kyoto is fully implemented by all parties with 100% compliance, and in the years afterwards, Annex I countries do not attempt to return to their old emission levels, but keep emissions to below baseline by the interval required for initial compliance. In the second scenario ("CONST"), not only is Kyoto implemented by all parties, but nations impose a tightening schedule of emission constraints so as to keep them frozen at, on average, 5.2% below 1990 levels through 2100. In the third scenario ("−1%"), participants go beyond CONST and actually reduce emissions below the Kyoto level by 1% per year, compounded, through 2100.

No defections, no leakages, no cheating. *And no effects*!

If Kyoto is fully implemented and the NOMORE scenario is observed, the concentration of CO_2 in the atmosphere reaches double its current level in about 105 years rather than 100 years. T-Rex rises by almost the same amounts as in the no-Kyoto case, with the average difference in 2100 only eight one-hundredths of a Bleep. Implementing the CONST scenario reduces T-Rex at 2100 by about twice that much. In other words, eliminating 100 years of emissions growth only changes the scenario by a further eight one-hundredths of a Bleep. The −1% scenario yields a total reduction of about 0.3 B below baseline.

If you think an increase in T-Rex of, say, +3 B is a big problem, then a reduction to +2.9 or even +2.7 is probably no comfort.

What is worse, the concessions negotiated at Bonn in July 2001 and Marrakech in November 2001 reduce the effects of full implementation even further. If the U.S. does not participate and the remaining signatories proceed with the deal as currently negotiated, the climate effect will be less than half of the original—that is, by 2100, T-Rex will have fallen about 0.04 B below baseline.

The clear message from these and similar projections is not that you should rely on 100-year climate forecasts but that Kyoto is of no consequence. If one believes in the models enough to conclude that global warming is a problem, it follows that the Kyoto Protocol is not a solution.

Environmentalists have lamented the inadequacy of the Kyoto agreement, pointing out that stabilizing global carbon dioxide *concentrations* (rather than emissions) will require a 70% reduction in emissions below 1990 levels. This equals a global cap of about two billion tonnes of carbon-equivalent (GtC), compared to current emissions of about 6.5 GtC.

There is no way to achieve such reductions without involving developing countries. In 1996, developing countries in Africa, South America, the Far East, Central Asia, China and the former Soviet Union together contributed about 47% of global carbon dioxide emissions (just over 3 GtC). And fossil fuel use outside the OECD is growing about three times faster than inside OECD countries, according to the International Energy Agency.

Even if the developed nations cut their emissions to zero, and thereby ceased being the developed nations, the developing countries would still have to cut their emissions by about one third.

The human toll that would be caused by these apocalyptic moves strikes some environmentalists as the price of a sort of Utopia.

Thinking About Costs, and Other Jobs for Grownups

So where are we? Pretend for the moment that the Kyoto Protocol assigns targets over quantities we can actually measure, and that participants will ignore the enormous incentives to defect, cheat and quit the agreement. Pretend that economic leakages won't undo half the effects of the treaty, that the models are all wrong and that Kyoto will actually slow global warming. (Oops. Go back a step. Pretend the models are right, just long enough to get worried about global warming so there is something to talk about, and *then* pretend they are wrong about Kyoto.)

We still need to confront the cost question: Is Kyoto worth it? Some analyses have suggested that the Kyoto Protocol could be implemented at little or no cost to the world's economy. This is untrue and is based on several fallacious lines of reasoning about economic policy changes. Before turning to some of the empirical cost estimates of the Kyoto Protocol, let's explore a few of the most common mistakes that have muddied the waters of the climate policy debate.

The "Beneficial Costs" Fallacy

In popular discussions of public policy, the distinction between costs and benefits is often confused. Suppose the government introduces a policy requiring all buildings to be painted pink. There would of course be a sudden surge in the demand for pink paint. Paint factories would gear up production, new staff would be hired and the house-painting industry would expand rapidly.

Some observers might consider the value of all this new employment and production as a *benefit* of the policy, but this is a misunderstanding. These are the *costs* of the policy. Homeowners are having to pay all these costs under the threat of prosecution, which means they would rather not do so.

The labour, materials and capital devoted to repainting all the buildings were taken out of useful service in the production of other goods that, had it not been for the regulation, the public would have preferred to receive. The cost of painting the buildings pink is the loss of all the other goods that might have been produced instead.

If there is no benefit to making all buildings pink, then the policy yields no benefits. The labour and materials used up in the painting process cannot be cited as a benefit, because those factors would have been employed elsewhere, producing goods and services that society actually wanted. The fallacy is to look at the private benefits of the painting industry and think of these as benefits to society in general.

The Free-Lunch Fallacy

Some environmentalists assert that regulations that force down carbon dioxide emissions will yield economic benefits by inducing firms and households to be more energy-efficient. One cannot base an assessment of the costs of a policy change on the assumption that there are unrealized gains to be found by forcing reductions in energy use. Economists call this a "free lunch" argument. If potential energy cost savings are so advantageous, we must assume that profit-maximizing firms and cost-conscious households would have implemented them already, or will soon do so. Energy efficiency measures that are only adopted under threat of prosecution can scarcely be assumed to be beneficial to the targeted firms; otherwise, the threats wouldn't be necessary.

In the absence of strong evidence to the contrary, we must assume that individual decisionmakers will take advantage of good opportunities to save money. Environmentalists who argue that firms are reluctant to reduce unit costs by improving energy efficiency are really saying that firms are not interested in maximizing profits. Under most other circumstances, this is not how environmentalists describe corporations.

The Fallacy of Irrelevant Comparisons

We should only compare the benefits of averting a disaster with the cost of a policy that would actually avert the disaster. One frequently hears claims like: "Yes, Kyoto will be costly. But when we consider the potentially devastating consequences of global warming, it is a small price to pay." Quite apart from the debatable claim about devastating consequences, the statement is wrong because Kyoto won't stop them. It is not relevant to hypothetical devastating consequences.

The "Precautionary Principle" Fallacy

The precautionary principle, as written in a 1992 UN treaty, states that, when confronting risk, uncertainty about the nature of the risk should not be a barrier to taking cost-effective preventive action. Others have interpreted it as meaning it is always better to prevent harm than to remedy it. It is essentially Item 7 of the Doctrine.

But life is not so simple. At any instant, one faces an array of risks. The roof over your head might collapse, for instance. You can either take steps to prevent it or just live with the risk and accept that you might need to remedy the problem later. To prevent the roof from falling, you could invest resources into reinforcing the building on the grounds that it is better to prevent the roof collapsing than to remedy the harm after the fact. But as soon as you pick up your tools to reinforce the roof, you are faced with on-the-job risks. And if you hire others to do the work, you face the risks that they will do a poor job and the workers face the risks associated with travel to the job site. According to the precautionary principle, you need to prevent all these risks too.

The precautionary principle in this form is vacuous, because it pertains to an artificial circumstance where there are two choices: one with risk and one without. This sort of circumstance doesn't occur in the real world. There are, typically, comparable risks from either action or inaction. The precautionary principle makes no sense in such a case. All it is—and all it has become in the climate change context—is a rationalization for one particular course of action that proponents prefer on entirely different grounds.

In the case of Kyoto, the precautionary principle is invoked to justify implementing it in the face of uncertainty because of the risks of global warming. But there is also a risk of death and hardship from the increased poverty that many nations will experience if Kyoto is implemented. The precautionary principle demands that we take cost-effective steps to avoid the risk of these impacts too. However, the only way to avoid the risks associated with diverting resources into compliance with Kyoto is to not put resources into compliance with Kyoto. So the precautionary principle counsels rejecting Kyoto too. So much for Item 7 of the Doctrine.

The Total/Marginal Fallacy

Any student of economics knows that decisions must be made based on *marginal* rather than *total* costs and benefits. This is a point often overlooked in public policy debates.

For instance, suppose a community is considering building a new school building. A referendum will be taken on whether the investment is worthwhile. One can imagine proponents of the investment using slogans such as "If you think education is expensive, try ignorance," or "You can't put a price on education." The fallacy is that these slogans refer to the total rather than the *marginal* value.

The real question is not whether the entire educational system is worth the entire expense but whether *one more* school is worth the marginal cost of providing it. The decision must be made at the margin.

If the marginal costs of a policy change exceed the marginal benefits, it means that the action is not socially beneficial. The costs reflect the value to society of the resources used up implementing the policy, while the benefits reflect the additional welfare created by the policy change. If the marginal costs are greater than the marginal benefits, people are being asked to give up resources worth more to them than the value of what is gained.

The analysis of policy changes in terms of marginal benefits and costs is explained in detail in any microeconomics textbook. It is sufficient to note here that, even when a policy can be shown to yield marginal benefits, this does not imply that it should be implemented. The marginal benefits must exceed the marginal costs. In the same way, showing a policy is costly does not imply it should *not* be implemented. For that conclusion, the marginal costs must exceed the marginal benefits.

The "Good First Step" Fallacy

It was pointed out previously that the emission reductions under the Kyoto Protocol will have no tangible effect on global warming, and hence will yield no benefits. Yet they will be very costly, as we will discuss in what follow. Numerous proponents have conceded that Kyoto itself will be largely useless, but argue that it is a "good first step" and that it should be seen as only the first of many such agreements, each one forcing emissions further and further down.

Eventually, so the reasoning goes, such policies will have the effect of slowing down global warming, at which point the benefits will appear.

The "good first step" fallacy is based on a failure to recognize that as emission cuts deepen, the marginal costs of such cuts rise quickly, while the marginal benefits do not. In environmental policy, the first step is always the one that costs the least and does the most good. After that, marginal costs of additional emission reductions rise and marginal benefits fall. If the first step fails the cost-benefit test, the subsequent ones will too; indeed, by a greater and greater margin.

That is to say, if the Kyoto Protocol fails the cost-benefit test, then *all* subsequent infrared gas emission reductions do as well, by a widening margin. If the cheapest emission reductions , i.e., the loss of those emitting activities with the least marginal value to society, generate more costs than the benefits associated with the first reductions in emissions, then further emission reductions will be even more disadvantageous. Therefore, if the Kyoto Protocol is a net loss to society, carbon dioxide emission reductions are unwarranted in whole or in part.

Of course there is always the dream of a promised land of new economics based on undiscovered science and technology to alter this basic pattern. But if there is such a place, since this is within the realm of fantasy, why couldn't we get there without any protocol and its risks?

Meeting the Kyoto Protocol Targets

If for some reason we do decide to try and meet the Kyoto targets, we ought to pick the strategy that costs the least. Finding a least-cost policy involves the "equimarginal criterion."

Picture the millions of points in a country where carbon dioxide is released, including every smokestack, house chimney, tailpipe, lawn mower and soda pop can. Suppose that at two of those release points, some efforts are being made to reduce emissions. At one place, emissions can be reduced at a cost of, say, $25 per tonne. At the other, the cost is $20 per tonne.

The reasons for the cost differences can include many things. For instance, they may be using different fuels, or they burn their fuels in different ways, or they use the energy for activities that are of different values. Regardless, the important point is that if you want to reduce emissions by 10 tonnes, it is cheaper to have the second emitter do all the emission reductions and the first emitter do none.

This says nothing about whether such an arrangement would be more fair or more politically acceptable, only that, arithmetically, 10 tonnes of emission reductions would cost $200 if the second emitter did all the cuts and $225 if the emission reduction targets were assigned equally to each source. And from the point of view of the atmosphere, it makes no difference where the emissions come from.

As a general rule, the more one source cuts its emissions back, the higher its marginal emission reduction costs go. So not only do we need to consider the current distribution of abatement costs at each source, but we need to recognize that those costs will change as emission levels vary. It will always be possible to rearrange emission reduction targets among the millions of different sources in a way that reduces the total dollar cost of hitting the reduction target, unless they are all operating at a point where further reductions would cost everyone the same amount. That's where the "equimarginal" term comes in. At the margin (i.e., considering the last unit of emission reduction), if the sources all face equal emission abatement costs, then the policy costs to the emitters have been minimized.

Most emission reduction policies being considered by countries like Canada do not satisfy the equimarginal criterion. They are not even close. One policy that is sure of satisfying it is a "carbon dioxide tax." It involves putting taxes on fossil fuels at rates that are in proportion to the carbon dioxide released when the fuel is burned. So anyone who buys fuel, of any kind, and burns it pays a charge that works out to a uniform rate per tonne of carbon dioxide emitted, even though the rate will vary across the dollar amounts of the different fuels.

In response to paying a fuel charge of this sort, people will soon start to think about how they can avoid paying the tax. Doing so leads to reduced emissions. The presence of a carbon dioxide tax will trigger millions of microeconomic decisions every day in which emitters compare the costs and benefits of certain courses of action. The costs will be things like forgoing profitable activities that would have involved using fuels, or in some cases substituting among different fuels when that is an option.

The private benefit of any of these actions will be the savings in the tax bill. The result of all these decisions will be that emissions start going down. Since everyone is paying the same tax rate on their emissions, they are all going to keep making emission reduction decisions up to the point where the cost of further action exceeds the benefit. And at that point, the marginal emission reductions will cost the same everywhere and will equal the carbon dioxide tax rate.

No governments have indicated a willingness to use carbon dioxide taxes. If the reason was that they were not going to cut carbon dioxide emissions, that would be fine. But governments are proposing instead all sorts of crazy emission reduction policies: subsidies for this and that, requirements for new technologies, changes to urban planning rules, tree planting, switching government-owned cars to natural gas, etc. It is highly unlikely this haphazard mix of rules and regulations satisfies the equimarginal criterion.

Why, then, have governments rejected the carbon dioxide tax? Perhaps the problem is that if a government were to implement one, it would have to admit to the public that the policy at hand is costly.

People would feel those costs in the form of the tax. Of course there is no reason the government has to keep the tax revenue. They could recycle it back into the economy by reducing other tax rates. And in some cases this may substantially reduce the economic impacts of the whole policy, though for emission cuts on the scale required for Kyoto it can never fully offset the costs. But whenever a tax is implemented, the government is saying, in effect, we are coercing you into supporting our chosen course of action. You have to pay the tax.

It is a reminder that governments do not simply dish out favours, they use up resources of the many to pay for favours to the few. Other policies have the political advantage that they emphasize the dishing-out-favours side of public life while burying the using-up-resources side. So if governments promise subsidies to owners of older homes to put insulation in their attics, they are playing up the dishing-out part while leaving the resource-using part of the policy in the shadows elsewhere.

But the economy responds in its own way to such policies and is not fooled by games of political optics. The economy registers the using up of resources, even if such losses are never officially acknowledged when governments make policy decisions. A new tax on carbon dioxide emissions would be politically costly. Leaving carbon dioxide emissions unregulated would also be politically costly. Yet, turning away from politics for a moment and looking at what would actually be good for society, leaving carbon dioxide unregulated would be the better option, and if we are going to reduce emissions, doing so with a carbon dioxide tax is the best option from an economic point of view. That being the case, what level should it be set at?

The Optimal Carbon Dioxide Price

In general, the appropriate level of a tax on an emitting substance is whatever the members of society decide is the value of the damages due to those emissions. Some people have pulled numbers out of the air for carbon dioxide, like $5 or $20 per tonne. But there are a couple of problems with just going with these cost numbers.

First, even if the damages really were worth $5, or even $20 per tonne—and we're not saying they are—the theme of emissions tax modelling in the last few years has been that the marginal social cost of carbon dioxide taxes is quite a bit higher than the tax rate itself! While the theory that pointed this out was developed in a little-known article by a Swedish economist named Agnar Sandmo back in 1977, the seriousness of the issue was brought home to economists only a few years ago when some economic modellers decided to test it out in a computational model.

The analysts[4] set up a dynamic model of the U.S. economy and worked in the assumption that CO_2 emissions generate damages of $5 per tonne. They found that a carbon dioxide tax of $5 per tonne would yield emission reductions of only about 4% but it would reduce real income by an amount that averages to $25 per tonne of emission reductions. The costs of emission reductions exceed the benefits for any target, however small. Going to higher emission reduction targets does not change the story.

Of course the above discussion begins with the assumption that we can meaningfully put a price on CO_2 emissions. But carbon dioxide is not a pollutant. It is not a health hazard, it does not affect visibility, it does not damage the paint on your car or emit a foul odour. It is as good for plants as oxygen is for folks. The only reason we might consider reducing emissions is the possibility that its infrared-absorbing characteristics have some role in the evolution of climate that some have conjectured to be deleterious.

And here we encounter the fundamental problem that derails any proposal to reduce CO_2 emissions: We are nescient as to the costs, if any, of a tonne of emissions of an infrared-absorptive gas like CO_2. First of all, we do not know the effect of CO_2 emissions on any particular version of T-Rex. T-Rex is an arbitrary construction based on tradition and expedience, rather than physical theory. No one experiences it, unless you count looking at a graph on a page as experience. If you do not like the way it is trending, just tilt the page slightly. Or pick another average over the temperature field that you like. There are lots.

You only ever experience local temperatures, local humidity, local precipitation, etc., and wherever you are, these things will change over the next few decades in ways that are fundamentally unpredictable. It is at the local level where the phenomenon of turbulence meets the decision-making criterion of economics and yields the answer to the question, what should we do about global warming?

Here is what we know. Of all the policy options open to us, using an emission tax is the least-cost approach. On the damages side, adding CO_2 to the atmosphere today may or may not lay a series of unpredictable local changes on top of an already unpredictable future sequence of local events. The changes may be somewhat benign or somewhat harmful. And there will be no way to know, before, during or after the fact, whether the changes have occurred and of what magnitude they were. Now ask yourself, how much are you willing to pay to put in place a policy that will have no measurable effect on any of this anyway? When you can answer this question, then you will know what price should be put on CO_2 emissions.

Meanwhile, Back at Kyoto

How much would Kyoto cost? To fix the discussion numerically, a central estimate of annual compliance costs for implementing Kyoto is 2% of national income. Is this a large amount? In 2010, the OECD will produce about US$25 trillion worth of goods and services. Two percent of this is $500 billion, about the same as the gross domestic product of Australia and New Zealand combined, and roughly seven times the entire foreign aid budget of developed countries. Over a decade, this amounts to $5 trillion.

This is a lot of money that must come out of real budgets and will therefore no longer be available for health care, education, highway maintenance, food inspection, homeless shelters, pollution control, wilderness preservation, foreign aid, support for the arts, or anything else for which it might have been used, including basic research into climate. All these things will be given up for a policy that even its strongest supporters admit will yield no significant change in the accumulation of carbon dioxide in the atmosphere.

Some defenders of global warming policy claim it will yield beneficial side-effects by reducing a host of other air contaminants related to fossil fuel use. But the need to reduce, say, sulphur dioxide emissions, or ground-level ozone, justifies policies which target these particular contaminants directly. It does not justify policies which are themselves pointless, but which might indirectly alleviate these.

Finally, there are distributional issues to address. Whether emissions are reduced by a carbon dioxide tax or some other means, it will disrupt household energy use. In the case of a tax, it will force up the price of home heating and other fuel uses. Since energy is a necessity, increases in its cost affect low-income households proportionately more, since they spend a proportionately larger fraction of their incomes on energy.

By implementing the Kyoto Protocol, not only will developed countries have fewer resources for dealing with real environmental problems, they will also have less money available for helping developing countries deal with their environmental problems, some of which are incomparably worse than anything the citizens of developed countries have to contend with. What less-developed countries need is growth and investment, to raise incomes and increase national wealth. As their incomes rise, resources become available for better environmental protection. The Kyoto Protocol will drain resources away in a futile attempt to prevent something no one really understands and wouldn't recognize even if it happened.

Fatally Flawed

The protocol is fatally flawed, as U.S. President Bush claimed, but not for the reasons he cited. The U.S. administration has repeatedly pointed to a lack of participation by developing countries as the deal breaker in the Kyoto process. This implies that if developing countries came on board, then Kyoto would become a good policy. But it would in fact make Kyoto an even worse policy than it already is, because it would spread the economic damage even further. The real problem is that the costs of emission reductions exceed any identifiable benefits. Pure mitigation strategies like Kyoto are an unwinnable gamble. If global warming is going to happen, a Kyoto-type agreement won't stop it, so it is a waste. If it isn't going to happen, then the policy is an even bigger waste. It is a net loss either way.

Is there any role for emissions reductions in climate change policy? At present, no. But to decide whether there ever will be a role for emission reductions requires constructing a new conceptual framework for connecting science to the policy process. This task is taken up in the final chapter.

1   Keith, David W. "Geoengineering." In *Encyclopedia of Global Change*, Oxford: Oxford University Press, 1998; National Academy of Sciences. *Policy Implications of Greenhouse Warming*. Washington, DC: National Academy Press, 1992.

2   *Washington Post*. "Research Casts Doubt on China's Pollution Claims." August 15, 2001, A16.

3   Wigley, T.M.L. "The Kyoto Protocol: CO_2, CH_4 and climate implications." *Geophysical Research Letters* 25(13), 1998, 2285–88.

4   Bovenberg, A. Lans and Lawrence H. Goulder. "Optimal Environmental Taxation in the Presence of Other Taxes: General-Equilibrium Analyses." *American Economic Review* 86(4), 1996, 985–1000.

10 After Doctrine: Making Policy Amid Uncertainty and Nescience

After Doctrine

In Chapter 1, we presented the Doctrine. To recap, it consists of the following familiar ideas.

1. The earth is warming.
2. Warming has already been observed.
3. Humans are causing it.
4. All but a handful of scientists on the fringe believe it.
5. Warming is bad.
6. Action is required immediately.
7. Any action is better than none.
8. Uncertainty only covers the ulterior motives of individuals aiming to stop needed action.
9. Those who defend uncertainty are bad people.

Let's go over these points one by one:

Is the earth warming?

After reading Chapters 4 and 5, this cannot seem like a very sensible question. It sets up a simplistic context in which to view climate change. No direct answer to the question is true; yes and no are both wrong. Someone who skipped straight to this chapter might be puzzled, but it is really a simple concept in the end. There is no single physical variable that describes warming and cooling, for the whole Earth; there is no single global temperature. This is just a basic truth of thermodynamics.

If there were some climate changes in the category of our sun going nova, or even something more moderate like a major ice age, then all of the infinity of local temperatures would be saying the same thing; namely, that it is heating or cooling everywhere and everywhen. But even the most strident and extreme scenarios put forward by the Panel of the United Nations (PUN) are Lilliputian in comparison. Contrary movements in the local temperature field everywhere would swamp any potential "signal."

The only way to force a simple warming or cooling picture onto a field of local temperatures is to say that, in some way, the places where and when it is cooling outweigh the places where and when it is warming. There is a problem doing this with temperature because there is no way to weight temperature at one point against temperature at another. Mathematicians would say it has no integration measure in space. It is an intensive thermodynamic variable, one of many quantities in thermodynamics that represents a condition and not an amount of something.

So if you want to take some average of the temperature field, it has very little physical meaning, unlike an average over energy, or the height of people in a classroom. Lots of—infinitely many—different averages over temperature are mathematically possible, but none of them has direct physical significance for the Earth. Any resulting average does not represent the temperature of anything at all. This is complicated by the huge variations in local temperatures up and down in comparison to what PUN claims to be looking for. You might get away with calling some average an index, but this huge variation virtually guarantees that different averages will behave quite differently from each other.

We have already seen that different published averages conducted over different pieces of the Earth's temperature field in different ways behave very differently from each other. This is to be expected. It would be remarkable if it were otherwise. Indeed, Professor Thermos, in Chapter 3, showed a simple example of a temperature field in which two equally plausible mathematical averages over the same data from a temperature field showed opposite trends. It would not be hard to construct families of averages over the Earth's data that decline instead of the more popular trend in the opposite direction. This is not physics; it is mathematics.

Plain-vanilla warming or cooling is not how climate change actually works. There is a lot more going on than temperature in climate change. And that is why the big climate models are not just thermometers, but must also treat a huge spectrum of fiendishly complex dynamics as best they can.

Has warming been observed already?

Sure. You can find lots of places in the world where temperatures have gone up recently. You can find lots of places where temperatures have gone down, too. But what people have in mind here is that an "unnatural" warming has been observed here or there. To conclude that would require some idea of what the natural temperature in a location is, but there is no such single thing. Nor do we have any theory of the climate that would define natural levels or rates of change in temperatures in any location.

Are humans causing it?

Causing what? Causing someone's favourite global average temperature statistic (i.e, T-Rex) to go up, or causing climate change? The former doesn't make much sense, while the latter is more complicated. Commonly used methods that purport to show humans causing climate change rely on climate models, which, as we showed in Chapter 3, are not suitable for this kind of purpose. And the absence of an underlying theory means that any statistical method, such as those talked about in chapters 5, 6 and 7, must fall short because they compare changes in some arbitrarily defined statistic to an arbitrarily defined alternative.

Moreover, as the signal is generated from a turbulent and chaotic deterministic dynamical process, we ultimately must be looking for a signal on an unknown chaotic carrier. In chaos cryptography, classic statistical methods cannot crack an encrypted signal carried on a chaotic carrier. The only way to do so is to have the equations of the chaotic carrier to work with. No one knows how to get hold of them for climate.

Do all but a handful of scientists on the fringe believe it?

This is the sort of claim that governments have found very useful to spread around. It certainly makes their job easier, since they can avoid the tough business of having to think through a complicated issue. As we discussed in Chapter 2, the interaction of government and Official Science has led to great declarations of certainty, as well as political marginalization of those who try to voice their doubts.

So it is understandable that people think only a handful of "marginal" critics have doubts. The critics are often referred to as "skeptics." A skeptic is someone that true believers do not want to invite to a seance. Neither are they generally invited to Big Panel meetings, though the two events are not quite the same.

One of the ways people have tried to argue that there is more than a handful of critics is to circulate petitions and get scientists to publicly own up to their doubts. We have not argued along this line. Instead we have simply shown, page after page, that certainty on the subject of the future direction of climate is impossible. There is no theoretical basis for predicting climate; models do not provide a substitute for theory, and the profound complexity of such things as turbulence and chaotic systems means that anyone who thinks we can predict the climate only courts the laughter of the gods. So if this part of the Doctrine is correct, then the scientific community is in a sorry state.

The truth of a scientific fact does not depend upon a democratic vote among "qualified" people. But that is what governments, Official Science and advocacy groups have pushed. Moreover, there is no consensus in the scientific community anyway. The existence of the PUN report does not establish that at all. It is like a collection of short stories on many topics, with a small summary attached in which a small group of people offers its interpretation. What each scientist wrote was neither conclusive nor comprehensive in terms of the overall position set out in the SPAMs. There was no simple proposition that the authors of the scientific report endorsed, and many would not agree with many aspects of the SPAMs.

There was no consensus among scientists, who are always debating debatable issues. And climate change is debatable, to say the least! What genuine scientists all agree on is that science is a personal journey. It is not an exercise in authoritarianism in which edicts on the truth sever the world into patriots and dissidents. Such a picture invokes aspects of politics offensive to free and thinking people. It is not science.

Is it bad? Should we act immediately?

Items 5 and 6 founder on the basic problem that if we cannot forecast something, we do not know whether it will be bad. Moreover, we cannot say it is bad or not if we aren't even sure what "it" is. Clearly we cannot say that "it" is all bad, and getting even more bad, as some seem to want to do. We cannot function rationally this way. It degenerates into nothing more than modern-day soothsaying without proper skepticism about the connections of bad events to climate change.

We may never even know if "it" is happening! As we discussed in Chapter 8, the genre of "impacts" studies is marked by an inordinate focus on temperature (often naively understood), a false belief that we can predict climate on a local scale far into the future, and a persistent inability to recognize the ingenuity with which people adapt to, and prosper under, changes of all kinds.

Is any action better than none?
Item 7 gets to the heart of thinking about policy. Chapter 9 talked about factors like costs and benefits and the question of whether Kyoto is worth doing. We think not. And there are many other related policy ideas that have been floated over the past few years that make equally little sense. Consequently item 7 cannot be true. Not every action is worth the expense.

Are critics of the Doctrine bad people with ulterior motives?
As for items 8 and 9, you are on your own with those. If you think the present book has been offered to the public as part of a plot by bad people who are just covering ulterior and malicious motives, well, so be it. Nothing we say now will convince you otherwise. And as we talked about in Chapter 7, uncertainty is not really the problem. Nescience is the problem. The difference between uncertainty and nescience is like the difference between looking for your lost puppy in your neighbourhood and looking for a purple dinosaur in an undefined parallel dimension.

Once that distinction between nescience and uncertainty is understood, we are in a position to have a realistic discussion about policy.

What Should We Do About Global Warming?
If you have skipped directly to this page, go back to the beginning of the book. This section will make no sense to you unless you have read what came before. It is certainly not a "Summary for Policymakers." Start at page 1 and take the time to read what follows. After all, a big part of the answer to the question above is that we ought to stop relying on quick summaries and actually take the time to think through this issue carefully.

The question as posed assumes that we are not already "doing" something when we are studying the issue. But at this point, study is precisely what we ought to be doing. There is a lot of science on which we can draw when trying to understand the climate change issue. But bear in mind a distinction: There is science that is good enough for, say, *Science* magazine, and there is science that is good enough for making policy. They are not the same things, even though they have been muddled together in this problem and many others involving big issues in science.

Articles that are good enough to publish do not have to be right or rock-solid or magisterial. They just have to be interesting to other scientists and free of any obvious errors, in the opinion of the editor and reviewers. Their contribution toward knowledge can be no more than that. An article like the one in which the hockey stick diagram (Chapter 5) was drawn fits these criteria. But that does not make it a sensible basis on which to change society.

The fact that an article was published does not relieve readers of the obligation to look critically upon its assertions and ask if they are really true or not, or to understand what the limitations are on the conclusions. Very few conclusions in such an article are free of a great many limitations and most authors go to great trouble to lay them out. However, there are also limitations implicit to a whole field that are not laid out. A reader is just supposed to know. It can happen that even some within a field lose track of these things.

There is no shortcut for decisionmakers that avoids the need to think carefully about the decisions they make. People who propose costly policy changes like the Kyoto Protocol cannot shield themselves from the responsibility attached to making such proposals simply by citing a published article or a scholarly authority. The peer review process provides a service to the scientific community for the purposes internal to science. This is not the same as providing a service to society for the purpose of ensuring policy decisions are sound. While we hope the two processes are not fundamentally contradictory, they are not the same. Peer review does not absolve policymakers of the responsibility for making bad decisions.

Nor should we suppose that we have climate change basically figured out, and all that remains is to shrink some bounds of uncertainty. The problem is much deeper. On the core issues we are confronted by an absence of knowledge. Under the circumstances, it is impossible to recommend a detailed plan of action when we do not know if action is even needed, nor what its effects would be. Therefore the only sensible recommendation is not to take action except for further study.

But study alone will not amount to progress if it continues along the same lines as those organized by the Big Panel. The fact that the Doctrine emerged on their watch, and that it has generated such deep public confusion and professional malaise over the global warming issue, is surely sufficient cause for us to consider a different way of studying the topic.

We will shortly propose a new method for public study of global warming, based on a different way of thinking about how science and public policy should be related. It requires that we shed some illusions about the easy presumption of certainty on complex topics, and the danger of applying political models of authority to scientific endeavours.

But first we must finish with the question in the heading. While familiar, it is badly posed. An improvement would be: *What, if anything, should we do about the fact that some scientists think adding carbon dioxide to the atmosphere will cause a deleterious change in the climate system?* The answer is that we should take the concern seriously enough to try and clarify the issue by posing some additional questions. How would we find out if this is true? Would it matter? Can we do anything about it? And would we want to, even if we could?

As to the first, the problem is that in the absence of a theory of climate, we would have no way of knowing if human-caused climate change were taking place, even if it were going on right before our eyes. Nor could we look back in time and determine that it had happened and that it was responsible for particular changes at any particular location, as opposed to natural long-term changes. We cannot expect to identify human-caused climate change even after the fact!

What we do know is that prosperous and free societies are best able to deal with changes, including adverse ones, regardless of cause. Therefore we ought to encourage freedom and prosperity around the world. All but a handful of scientists on the fringe believe this.

On the second question, we do not know how this or that particular average of temperatures ought to behave in the absence of fossil fuel use, so whether T-Rex is rising or falling means nothing in terms of interpreting our influence on the world. And T-Rex only exists on paper anyway. We experience an infinite number of temperatures at every location on

Earth, not an average cobbled together from white boxes and good intentions. If things are changing in the climate, we have to look at them directly in all their particularity and complexity, rather than trying to reduce the problem to the artificial simplicity of a one-dimensional character in a B-movie.

As to the third question, the climate is not a clock we can take apart and put back together or manipulate at will. Even if the world were as sensitive to CO_2 as the Big Panel has tried to argue, all the king's horses and all the king's men have not succeeded in coming up with anything better than the Kyoto Protocol, which, with its compromises at Bonn and prostrations at Marrakech, make it pointless with regard to any effect on atmospheric carbon dioxide concentrations. T-Rex will do what it was going to do despite it.

The world is full of surprises. "Surprises" was precisely the word used to describe it in Section 14.2.2 in PUN's scientific report. But too few actually read the actual scientific report. There the real scientists were able to send us a message in a bottle through the sea of politics to tell us that they too do not know. For all we know, we may yet be on the verge of an ice age, in which case we won't think global warming such a bad thing.

No one knows why some versions of T-Rex are rising and others are not. Temperatures rise and fall in response to changes in solar output, fluctuations in the earth's orbit and the lunar pull on tides, volcanic activity, long-term ocean dynamics and many other factors we do not understand, including every flap of a seagull's wings. And sometimes it rises and falls for no external reason at all, just because nonlinear, turbulent and chaotic dynamics can generate large sudden uncaused changes in the state of a system. That is what they said in 14.2.2 and that is what we have shown in this book. The presumption that every marked change has a cause is simply wrong. On Earth, these things are beyond our understanding, let alone our control.

As to the question of whether we would *want* to change the climate even if we could, we must look at the costs and benefits of small steps. The steps that involve carbon dioxide emission reductions (such as Kyoto) cost more than any benefits they are likely to generate. So they are not worth taking.

That said, some kinds of economic changes, driven by the internal comparison of costs and benefits, lead to emission reductions. For instance, there is a continuing drive toward energy efficiency in industrial economies. Every year we get a bit more efficient in how we use fuels, because it is economical to do so. But we cannot point to those actions and say that *therefore* all carbon dioxide emission reduction proposals are warranted. The question is not whether beneficial emission reductions are warranted: They will be undertaken with or without a push from government. The question is whether costly emission reductions are warranted, and the answer is no.

Even so, some people will rebel against this counsel. Over the past couple of years, as we have talked to people in and out of government, we keep hearing the message, *Yes but we have to do something.* Whether the rationale is the "precautionary principle" (which doesn't apply in this case) or some other dressing up of a simple gut reaction, there is a prevailing assumption out there that it is unacceptable to simply "do nothing."

At some point, we hit the wall in terms of what reason can accomplish. For those who really feel that way, despite all we have argued, we have run out of arguments. We could keep rephrasing what we have already written, but we won't. What is needed is for the reader to climb over the psychological hurdle of admitting that we should not burden society with carbon dioxide emission reduction requirements.

Regardless of how much effort and ingenuity has been expended developing the Kyoto Accord, it is the wrong thing to do. The right thing to do is to muddle along, focusing on basic priorities like economic development, wealth creation, education and the spread of freedom. We know these are good things to do. Policies like Kyoto do nobody any good and only take resources away from those real priorities.

So the best policy on global warming is to make sure science is free to investigate it, without having to prove constantly that this or that is relevant to policy issues. This has created a climate in the research community that has systematically pushed out the absolutely essential basic research that is all but disappearing because of it. Adequate funds must be made available to those who are studying it just because the research is good, not because it is producing fodder for political action.

Otherwise, the best policy is to do nothing unless future information indicates otherwise. This has been regarded as a political position. "More study" does not appeal to impatient political ideologues. But it is the only course that reflects a proper humility in the face of the enormity of the scientific problem we are facing.

The Law of Group Polarization

How can we ensure that science is truly free to investigate the question of climate? The first challenge is to avoid the sort of convection of certainty that we have observed with global warming, discussed in chapters 1 and 2. There we discussed a feedback between Official Science and the politicians who consult them, amplified by the interventions of the media and activists, leading to ever-increasing declarations of certainty concerning global warming even where certainty is inherently impossible. In the succeeding chapters, we have explored how the Doctrine of Certainty sits at odds with reality, leading to the international pursuit of costly, unattainable and pointless policy goals.

The problem cannot be reduced to a mere quibble about whether the Big Panel report lacks balance. After all, the report is pretty frank about the problems, if you know where to look. For instance, the text of the assessment report (p. 570) says: "There are still severe limitations in the ability of [climate] models to represent the full complexity of observed variability..."; "Some new results have challenged the conclusions drawn in earlier reports..."; and "There are areas where there is no clear indications of possible changes or no consensus on model predictions..." If you understand the jargon, you'll also see that Section 14.2.2 of the report, mentioned above, says we don't have a clue how the climate is going to behave in the future!

Of course there is a legitimate charge to be made here that the SPAM lacks balance. All these caveats are summarized in the SPAM as: "Confidence in the ability of [climate] models to provide useful projections of future climate has improved due to their demonstrated performance on a range of space and time scales." To say the least, this hardly does justice to the report. But it encapsulates quite neatly the startling fact that Official Science could look over a landscape of uncertainty and nescience and report back that we pretty much know what the problem is and how to fix it.

Things might have gone differently. Why, for instance, did the social "convection" process, discussed in Chapter 2, lead to an unwarranted degree of *certainty*, rather than, say, an unwarranted dismissal of the whole issue? That is, can we say anything about why the Big Panel tilted in the direction it did, and was that outcome inevitable?

Chapter 2 dealt with this convection process and why the different groups involved naturally tilt in the direction they do. But here we are asking about how individuals form groups that have a tilt. In recent years, social psychologists have begun to gain some insight into how groups deliberate. We have long heard the notion that group deliberation ought to produce better, wiser judgements than people are capable of forming on their own.

Surprisingly, though, what happens is often the opposite. People who enter a group deliberative process have a certain view of a subject, but the deliberation may push them to a more extreme version of the view they originally held, and the view of the group then becomes heavily tilted in one direction. Not long ago, a law professor at the University of Chicago, Cass Sunstein, published an essay on this phenomenon called "The Law of Group Polarization."[1]

Examples are easy to find. Suppose some university professors who are inclined to support gun control get together to discuss a proposed law. We would not be surprised if they end up supporting a declaration more adamantly in favour of gun control than any one of them would have produced as individuals. Suppose a group of citizens who are somewhat inclined to support military action against drug smugglers meet on the Internet. We could easily believe that, after discussion, they end up promoting more extreme views on the subject than when they began. And so forth.

Wherever groups form to discuss political, religious, legal, economic or other kinds of issues, the group members often move toward an extreme version of the position that was a mere tendency prior to deliberation. Experiments by social psychologists have confirmed this pattern in all kinds of settings. It certainly arises when complex moral and political questions are on the table, but it happens even when a group is convened to discuss obscure points of fact like the height of a particular city above sea level.

Many aspects of psychology come into play in this process. People defer to those they consider experts. Even among experts, there are people regarded as being more and less expert point by point. People want to be well regarded by their peers, and especially those they consider superior. They modify their positions based on their perception of what everyone else believes. Some like to define themselves as occupying a more radical position in comparison to others, whereas others like to play the moderate; they attend to the degree of confidence and inflexibility exhibited by others, etc. All these details are explored by Sunstein and the authors he reviews.

Group polarization doesn't mean an approach to the middle ground. It means instead that deliberation within a group can produce more extreme views in one direction than were held previously. As a result, the median view of the group is itself radicalized.

Among the factors that can cause polarization, two are particularly pertinent for our inquiry. First, group self-selection is a factor. Groups in which members are brought together out of a common interest will tend to have a more homogeneous character and the deliberation process will involve exposure to a more limited range of arguments. People within homogeneous groups will hear their own views echoed back to themselves more frequently, pushing them further along in the same direction.

Second, the requirement that the *group itself take a position* causes movement toward the within-group extreme. Sunstein comments that "the requirement of unanimity may well, for example, produce a shift toward the most extreme points."

Both these factors apply to the Big Panel. As to self-selection in Working Group I, there are 118 coordinating lead authors or lead authors, not counting PUN co-chairs. Of these, nearly half (57) work at a handful of government-funded labs. This is not a representative sample even of climate scientists. This small group is to be contrasted with the 1,000 authors and referees who participated in the project as a whole. The significance of the small group is its importance in shaping the interpretations and final form of the scientific report written by the others.

It would not be difficult to empanel an alternative selection of credible scientists who would interpret the same scientific material somewhat differently than Working Group I did. This was not reflected in the various summaries. The blueprint of the process called for consensus and unanimity, so we got exactly that.

People who do not share that point of view either stay away because they don't care for the hassle, or they join in the process because they feel a duty to provide balance. But they have to have thick skin, because they are commonly held up as contrarians, skeptics, dissidents and malcontents.

A lot of attention gets paid to scientists like M.I.T. Meteorlogist Richard Lindzen, because he is one of the few who vocally objects to the so-called majority view. Rather than taking this as a signal of what his colleagues may also be thinking, his dissent is viewed within Official Science as a psychological curiosity, and he is sometimes held up like a specimen from Ripley's Believe It or Not museum.

It will take a generation to unravel this mess. Back in 1994, an atmospheric physicist at Penn State, Craig Bohren, made a speech at his university in which he lamented what the global warming issue was doing to his profession. His charges were blunt:

> The government's response to clamoring from an electorate frightened by global warmers to do something about global warming is to recklessly toss money to the wind, where it is eagerly grasped by various opportunists and porch-climbers.... I have never understood Gresham's law in economics—bad money drives good money out of circulation—but I do understand this law applied to science. Incompetent, dishonest, opportunistic, porch-climbing scientists will provide certainty where none exists, thereby driving out of circulation those scientists who can only confess to honest ignorance and uncertainty.

It will take many years before graduate students in many disciplines are free to question the Doctrine, to scrutinize and question the data they are working with, to treat GCMs as mere models rather than ECMs (Enchanted Computing Machines) and generally to act like scientists again rather than recruits in a cause.

The essential problem of the Big Panel is the requirement for it to have a consensus. As mentioned above, this enhances the push to group polarization. Requiring a consensus or a "group" view raises the stakes for those who want to argue against the existing tilt. It is not enough just to get your view forward, you have to carry the whole group or there is little point to making the effort.

On many university campuses, the student unions and newspapers often take political positions that the majority of students find ridiculous and even bizarre. Yet year after year, decade after decade, university student councils remain alienated from the mainstream of campus opinion. The problem is that no individual student has an incentive to get elected and try to change the views of an entrenched council. The only people attracted into these societies share the odd views of the incumbents and serve to perpetuate them. Similarly, a lone scientist who volunteers to serve as a lead author for the Big Panel with the intent of pulling it away from the Doctrine faces an almost hopeless task. So why bother?

Professor Sunstein rightly does not assume that group polarization is automatically a bad thing. If a group deliberates, and in the process moves to a more extreme view, that may be appropriate, given the circumstances and evidence available. But the regularity with which this outcome occurs does give us cause to be concerned that institutions for deliberating complex issues can be manipulated.

If the group is selected with a bit of a tilt, that tilt may lead to a big slant over time. In that case, the outcome was predetermined by selecting a few principals with the preferred outlook. In many cases, policy questions must be made based on complicated and uncertain science. Given the nature of group polarization, appointing an "expert panel" can amount to a way of dressing up a prior conclusion with some imperial robes.

But there is a way around this problem.

In Praise of Polarization: The People v. Carbon Dioxide

In practical terms, we cannot avoid the creation of something like Official Science. There will probably always be a layer of personnel standing between regular science and politicians, although politicians can help things by doing more of their homework. But if we must have Official Science, we can at least rethink, very carefully, what we are trying to achieve, and how to avoid triggering certainty convection.

When important but complex questions are under public investigation, there are two kinds of inquiries that can be used: *fact-finding* or *adjudication*. The first type of inquiry is relatively easy. It involves collecting information, stapling it at the top left corner and handing it in. This is appropriate when there is agreement on the scope of the factual material and how it should be collected, and its meaning is self-evident.

How does France finance its health care system? Send a fact-finding team. How many teachers are going to retire in the next decade? Appoint a fact-finding committee. In cases like this, group polarization is much less of a problem, especially if there is agreement about the sources of authoritative information and the only task is to look it all up.

The second kind of inquiry, adjudication, is much more difficult. Its task is not merely to gather facts but to interpret them. This is where scientific disputes can emerge that ought to be allowed to run their courses without tilting the field in favour of one outcome or another because of exterior policy demands. However, this is the situation in which group polarization is most likely to occur.

In every other area of society, when a task requires adjudication—that is, a judgement as to the meaning of the available data by someone in a position of authority—no one would think of using a fact-finding model. Instead we turn to one *in which contrasting opinions are deliberately sought out and given a full and fair hearing.*

Courts insist upon competent representation for both the prosecutor and the defendant, and will suspend proceedings if one or the other is missing. Each side is given all the time it needs to present its case. The testimony of each witness is cross-examined. Each side can bring in its own experts. Attention must focus on the facts of the matter and the logic of the case, and not on the character or motivation of those presenting arguments. If the losing side can show that the court displayed prejudice, another court is asked to start over and re-try the case.

We cannot be naive about the human propensity for group polarization. Whether the subject under dispute is the influence of infrared gases on weather formation, fiscal-policy reform or the best way to make a pie crust, people approach their subject with pre-existing commitments, preferences and notions of what is and is not plausible. Groups that organize themselves to advance a common position will in the process tend to become more extreme in that position. The only way to make sure that this tendency does not render the whole policymaking process dysfunctional is to ensure that more than one group is formed, and each one is charged from the outset with producing a report that advances a particular position.

Let us emphasize this point. When a controversial decision at the interface of science and policy must be made, politicians too often look for advice from a mythical creature called the "neutral" scientist. If it is a pure fact-finding mission, that is probably all right. But many policy decisions are not based on mere fact-finding, they are based on fact-*interpreting*.

So note well: Scientists have opinions too. They are human beings. They go to work each day carrying life histories of experiences and thoughts and aspirations and passions that are intertwined with their chosen work. They will be predisposed to regard some things as plausible and other things as implausible, sometimes based on nothing more solid than a gut feeling or some long-forgotten remark their third-grade teacher make.

If their views become the topic of scientific inquiry, and they are properly trained, they will test them relentlessly against data and theory and common sense. But these predispositions can sometimes create limits on the range and quality of questions an individual scientist asks. And the kinds of biases that creep in will differ from one scientist to another. That's why they need each other. That's why they need the culture of science to do precisely what the Big Panel has not done: deliberately cultivate, and hear, *opposing* points of view!

Look at any of the myriad issues where uncertain science and hot policy decisions are linked. Should cities ban lawn pesticides? What is the effect of abortion on women's health? Is there a link between violence on TV and adolescent delinquency? Can homosexuals go straight? Do cellphones cause cancer? Do above-ground nuclear tests pose a health threat? How many bird species are at risk of extinction? Should we bring back DDT for controlling mosquitoes? Does air pollution cause asthma? It goes on and on.

In each of these issues, and in many more, the usual sequence of events for finding answers goes something like this. The government wants advice on the science. It looks for the "neutral" expert. By some chain of rumour, acquaintance and political jockeying, Dr. Bland is selected to form a panel and write a report. The panel polarizes around a view. The Bland Report comes out, 500 pages long, dense with footnotes. It all boils down to a conclusion, which as it happens was precisely the view that Dr. Bland and the other panellists held before writing the report.

Then some other people start to object. They say they weren't consulted, or that the Bland Panel overlooked some important evidence. But by now the government has institutionalized the Bland Report. The opponents are the "skeptics," the minority, the outsiders. It doesn't matter how many of them there are, how big are the errors they find in the Bland Report or how good their own arguments are. They do not have the money or the institutional credibility to produce a report of their own.

When they speak up, they do so as individuals, and they can never carry the weight or *gravitas* of the Official Bland Panel. Some resort to publishing an occasional op-ed or setting up a Web site. At the same time, they resent being forced into that venue. Most just get frustrated and drop out of the debate. Their expertise gets lost just when we need it most. Thus it is that big policy decisions get made, time and again, on the basis of incomplete and unbalanced science.

If the PUN's only job was to collect data, there might not be a problem. But the PUN also offers an interpretation of the data, without any guidance from a proper theory or even general agreement on which data to use. If it is to be taken seriously as an adjudicative body, it must be set up as one. That is, it must act as an impartial judge, not as the lead prosecutor in the case of *The People v. Carbon Dioxide* (or, for those of us under the British system, R. *v.* CO_2). The fact that the PUN has set itself up as the prosecutor of one side in an uncertain debate means it should not be adjudicating this issue, any more than a prosecuting attorney could simultaneously sit on the bench.

In gathering a small number of like-minded principals to run the PUN, the organizers (at the UN) assumed falsely that these scientists are not subjective, that they do not have beliefs and presuppositions that tilt their views. This is not a special flaw in the PUN, it is merely what happens when humans process large amounts of information on uncertain subjects with weighty implications. The exercise is always undertaken in the context of a world view.

We do not need to guess what is the world view of the Big Panel leaders. They do not attempt to hide it. They are committed, heart and soul, to the Doctrine. They believe it and they are advocates on its behalf. They have assembled a body of evidence that they feel supports it and they travel the world promoting it.

There would be nothing wrong with this if it were only one half of a larger exercise in adjudication. But governments around the world have made the staggering error of treating the PUN as if it is the only side we should listen to in the adjudication process. What is worse, when on a regular basis other scientists and scholars stand up and publicly disagree with the Big Panel, governments panic because they are afraid the issue will get complicated and undermine the sense of certainty that justifies their policy choices. So they label alternative views "marginal" and those who hold them "dissidents."

If we set up courts this way, it would mean the prosecuting attorney presents his case, announces that anyone who would speak on behalf of the defendant is an enemy of society, then climbs on the bench and delivers a verdict, without having allowed the defence to address the court. In other words, if we ran courts the way the Big Panel works, we would call them show trials.

This book should not be taken as the case for the defence, but for allowing a defence to be presented, according to sensible procedures.

Should cities ban lawn pesticides? We do not know. But we know how they ought to decide. A city should form two panels. One is asked to produce the strongest case possible for banning them. The other is asked to produce the strongest case possible for their use. Then each team gets to write a rebuttal to the other's. The final report consists of all four documents, without a summary.

Two teams? Handpicked so they hold foregone conclusions? Sure. Let them polarize all they like. Let them self-select their members and tilt into their preferred position. In the end, their reports will be set side by side. If they are evenly matched, so be it. That is the honest message of the science. And any process that fails to convey it is perpetrating a fraud on the public.

In the case of climate change, we would begin by recognizing that there are opposing views, and it is not obvious from the outset which is correct on any particular question. We would form two groups with equal funding and adequate membership in each. One group would be called Heads and the other group Tails. The job of the Heads group would be to produce a report making as strong a case as possible that human activity is causing a huge climate change that will have harmful consequences. The Tails group would have the job of making as strong a case as possible to the contrary.

Those who have been hanging around the climate change field for the past decade would have no difficulty identifying who should be on the Heads team and who should be on the Tails team. And since we would have done away with the artificial labels of "mainstream" and "marginal," a wider range of participants would come forward, especially for what today is maligned as the "skeptical" side.

Each group would be asked to produce a 200-page report, as well as a 50-page rebuttal to the other group. The complete 500-page document would be released without a summary, but with an index. It would be submitted to the world's governments without either panel being asked to render a decision on which team's report is stronger.

Each government would have to decide for itself. Ministers would not have summaries to give them phony detours around the science. Politicians would have to knuckle down and read the actual reports. They could, if they like, consult internal and external experts for their opinions. But even if one government made the mistake of setting up a national Official Science group to render a verdict and write a SPAM, it would not be binding on any other country. As no country would want to jump to a conclusion that its allies all contradicted, there would be an incentive for cabinets to really understand the material so as to put policy decisions on the securest possible footing.

We can imagine the protests that supporters of the Big Panel would raise against this sort of approach. "It would lead to confusion," they will say. Why? "Because the report would not render a bottom-line decision. Everyone will conclude what they want from it."

Oh really? Does the Big Panel fear that the Heads and Tails reports would be so evenly matched it would not be obvious which is the stronger case? That would seem to be an admission that the position espoused in recent years by the PUN is not nearly as conclusive as it has been claiming. But if they do think it will be obvious which is the stronger report, then what's the problem? If the Heads group really have such a strong case, putting it alongside the Tails case will only sharpen the contrast, especially since the Heads group gets to produce a rebuttal.

We propose this as a replacement for Working Group I. Working groups II and III should simply be shut down. These groups interpret the significance of the conclusions about climate science for society. By 1995, they had already said everything that needed to be said on the subject, until such time as WGI has "finished" and produced definitive findings.

The fact that WGI continues to operate means that there isn't the information base on which to give WGII and III a coherent mandate. Moreover, they are trying to make policy recommendations that sovereign states must make, not UN committees. Individual countries can set up their own Little Panels to look at adaptation and mitigation policy, but the reports from working groups II and III have no bearing on national policymaking.

Look at the policies being proposed by the governments considering Kyoto implementation. Not one of them has used the reports of working Groups II and III as the foundation for their own policies. If the governments who asked for the reports don't bother following their advice, why should the rest of us continue to pay for them?

The Flying Dutchman

It is once again a dark and stormy afternoon. The winds howl and the dog cowers, while thunder rolls and lightning flashes. The thunderstorms of summer are back, wild and wonderful as ever. Humans have the ability to forget things like the zip up the spine that a crack of thunder can deliver. But we are reminded of the immense power of the storm in the rumble of its voice—power enough to flatten homes and wreak havoc. The dog is smart to cower—but he misses the joy and wonder of it.

Power so great, but it is too small to show up in the best computer climate models. Our busy lives are often put on hold while the reality of the storm passes, yet many do not call it reality. Reality in the minds of many is our puny political operas in which a thunderstorm is just a prop. The most strident proponents of the Doctrine are like *The Flying Dutchman*, sinners doomed on a stormy sea in search of redemption. Their vision of the world is a humourless dark storm, heavy with brooding and ever so Wagnerian.

In the preface, we listed our reasons for writing this book. There is one more that we did not mention: We wrote it for fun. Of course it was a lot of work and more than a little strain and hassle at times. Nonetheless, it was a chance for us to say things we have wanted to say for a long time. The tone of the public discourse on this subject has become too Wagnerian, too pompous and self-important for our taste. And for academics to complain about that is really saying something.

Life is too short to get overwrought about the panic of the day. We met people along the way who are consumed with the politics of global warming, and whether they are for or against Kyoto, it seems to have drained all pleasure out of their lives. While these can be absorbing policy issues, ultimately they are not that important. What is important is laughing over meals with friends and family or talking over drinks and enjoying interesting things. Whatever the climate does, and whatever the nations of the world decide about Kyoto, these things will still happen and will still help mark out what is good about life, despite the dubious apocalyptic prognostications.

We thought global warming important enough to write a book about it. But we also want to tell you that it is not all-important. It merits attention and careful thought, but not operatic melodrama.

There were times when it would have been easy to lose sight of these things ourselves, so absorbing is the storm we were studying. Big life events sometimes helped us refocus. During the writing of this book, one of us was successfully treated for cancer. And there were little things. One of us saw his youngest child enter preschool while one saw his youngest child finish high school. It is easy to get caught up in the storm of public issues and forget what all politics, laws, armies, economics and science are ultimately for.

Every human-made thing around us, every sophisticated idea, is the gift of the generations. It is the legacy of people knowing. We all contribute to it in one manner or another, either by personal contact with people in our lives or through more formal means. It all accrues to make the legacy of the next generation wealthier than the previous one.

But passing this legacy on can be extremely difficult. There has been a loss of nerve in passing the torch between the generations. While we cherish what we have inherited from those who went before us, there is a cost in taking it up. The basic homework it entails is so much harder than the political opera.

It is extraordinary that even sophisticated and educated people of our heavily scientific age are so skittish about technical details that they have to be provided in the form of children's picture books.

Not only is essential mathematical symbolism not desired; it is regarded as provocative. It is taught in every high school in the world, but we were repeatedly cautioned while writing this book not to use equations or to dwell on scientific technicalities, *even* when they are at the heart of the issue. Talking about this issue's mathematics is a bit like attempting to do sex education in Victorian England. It's something that people need to hear, but you can't talk too directly about it, and people squirm the whole time.

On scientific matters such as global warming, the political sophistication far exceeds the existing scientific understanding. They are wildly out of sync with each other. We cannot cope wisely with complex technical issues while this is so. We will always get caught up wrangling about the politics and the social science before the basic scientific issues are even on the table.

We have to learn to have more respect for the technicalities. On important issues we are not already familiar with, we cannot live on executive summaries. Unfortunately, there are few good mechanisms for public education on important scientific issues. The newspapers, as they are constituted, are incapable of fulfilling that function. They provide science junk food at their best.

Some people talk a lot about our responsibility not to abuse the natural world around us, but, while living in a technological world, this means they must also assume the responsibility for the knowledge of their world. But we have seen that this is not done. That it isn't is as much at the heart of why we have a political storm over climate change science as anything else. They rely unwisely on authoritarian pronouncements rather than thinking with their own heads.

Even some scientists themselves are not comfortable with anything other than the tried and true methods of their respective specialties. Moreover, the full PUN report itself is quite lean on technical details and long on commentary, uncontroversial though most of it is—there are few equations even in its "scientific" document. This reflects the state of our global society, and its limits in dealing with scientific issues, of which global warming is only one of many.

The previous generations left us a legacy we too often take for granted. And even it does not begin to answer the wonder of it all. G.K. Chesterton once said the real reason water flows downhill is that it is bewitched. He was reminding us not to lose the sense of wonder we felt as children, either by flattering ourselves that we have the whole world figured out or because we cower in fear and are afraid to look.

The storm overhead is an intimidating thing, but it and the wide world of which it is a part invites us nonetheless to take courage and explore. It requires that we be willing to think deeply and clearly about complex things, and not to let shapeless fears and phantom threats hold us back.

The thunderstorm has passed, leaving clear air in its wake. The sun sets and the summer stars blaze across the sky. There is the summer triangle, and running through it is the Milky Way at its best.

Your grandparents, and the generations before them, saw exactly the same stars in the same places after similar stormy summer days during the centuries past. All the generations, back to the beginning, stood under the same sublime heavens and wondered what the future would hold for themselves and generations to come. They wondered about us.

It's our turn to wonder about the future and the generations to come. There are great perils and wonders facing humanity. That is not new.

Humanity has always faced them. They are here on Earth and out there among the stars. We do not need to imagine them or conjure them up. They are out there, waiting for us. Sooner or later we will face them.

We will need our wits about us. We will need what we have learned from all the ages past. If we cherish it, all the generations back to the beginning will be with us at every step. The stars will be filled with possibilities. No need to be afraid. We will learn. That is our redemption.

---

[1] Sunstein, Cass. "The Law of Group Polarization." University of Chicago: John M. Olin Law and Economics Working Paper, Series 2, No 91, 1999. www.law.uchicago.edu/Publications/Working/index.html.

Glossary

Average   Any one of an infinite number of possible values that can be used as a representative of a given collection of numbers.

Aerosols   Microscopic particles, formed of every known substance, that get carried by air movements into the atmosphere and slowly fall to the surface.

Ambient heat prejudice   The tendency to assume that increased infrared absorption by atmospheric gases must raise the air temperature.

Anomaly   The difference between a data point and a mean.

Autocorrelation   A statistic that measures how the value of a series relates to values at other times.

Big Panel   See PUN.

Chaos   The property of solutions of certain systems of differential equations in which the solution path is sensitive to small changes in initial conditions, yet remains bounded.

Climate   Loosely defined as the "average" or prevailing weather in a region, over some span of time.

Convection   The motion of fluids in the atmosphere involving heating from below and cooling from above.

Doctrine of Certainty   The claim that we understand the climate, what is wrong with it and how to fix it, and that one ought not to say otherwise publicly. The specific tenets as they apply to the global warming issue are listed in Chapter 1.

Extensive quantity   In thermodynamics, a quantity that occurs in amounts that can be added up.

Fractal   A geometric shape that does not smooth out no matter how much you magnify a surface. Mathematically, it may have a fractional dimension.

Function   A rule for relating two or more variables.

Global warming   The hypothetical situation in which the temperature field changes in some way that involves general increases in temperature around the world.

Greenhouse effect   Term applied to describe the action of infrared-absorbing gases in the atmosphere, though the process in that case is quite different (see Chapter 3). In greenhouses, energy drain from turbulent fluid dynamics is impeded by a physical barrier, requiring increased radiative drain and a higher air temperature. There is no such impedance to turbulent motions in the atmosphere.

Gridding   Grouping temperature observations into geographical regions to compensate for the uneven spread of weather stations.

Infrared   The portion of the radiation spectrum that is too "red" for our eyes to see, but not as red as microwaves.

Infrared-absorbing gases   Gases that absorb specific bands of infrared energy and re-emit it to their surroundings. These are often misleadingly called greenhouse gases.

Intensive variable   In thermodynamics, a quantity that does not exist in amounts, but rather describes a condition of a system, such as temperature.

Kyoto Protocol   A 1997 treaty that requires developed nations to reduce emissions of infrared-absorbing gases.

Latent heat   Energy in gaseous water that is released when it turns liquid.

Leakages   Increases in emissions in one region induced by policies to reduce emissions in another region.

Linear regression   A statistical method that maps one variable onto another (or a group of variables) allowing for random residuals to cover gaps between the fitted values and the actual values.

Lorenz equations   A dynamical system describing atmospheric convection in which the phenomenon of chaos was first encountered.

Map   In mathematics, it is a rule that determines which values in one set get taken to which values in another.

Mean   A popular type of average. The sum of a group of numbers divided by the number of numbers in the group.

Median   A type of average. In a collection of numbers, there is a range of numbers that have as many members of the collection above it as below. The median is in the middle of that range. It is normally different in value than the mean.

Model   In physical science applications, a set of algebraic equations used to describe a natural phenomenon. The equations may correspond to those of a physical theory or they may be approximations devised for computational purposes.

Official Science   The body that serves as an interface between the practitioners engaged in the personal experience of scientific research and policymakers.

Paleoclimatology   The study of climates at times before systematic collection of weather data.

Parameterization   An algebraic approximation to behaviour that is governed by equations that are not known or not useable for computation.

Probability density function   A curve that describes the behaviour of random variables following a known distribution, by showing the relative occurrence of data points across a range of values.

PUN   The Panel of the United Nations that produces reports about climate change. (It is also known as the Intergovernmental Panel on Climate Change.)

Radiative transfer theory   The equations that govern the drain of energy from a surface by radiation emission.

Stationarity   A property of a random variable in which the mean and variance are finite and do not change over time. In the case of mappings, the property that the coefficients do not change over time.

Statistical significance   A property of an estimated mapping parameter. It implies that if it were actually zero, the data would have yielded a value at least as large as the one observed less than five percent of the time.

Stratosphere   The region of the atmosphere above the troposphere, approximately 15 km and higher, to the edge of space.

SPAM   The summary for policy makers and media produced in 2001 by the PUN.

Sub-grid scale processes   Phenomena that occur on too small a scale (spatial or time) to be individually computed by a model.

Theory   In physical science, a precise set of equations and principles that govern the behaviour of a natural system.

Troposphere   The lower part of the atmosphere, from the surface up to the stratosphere.

Turbulence   The nonlinear motion of fluids involving rapid eddies, whorls and vortices.

Variable   A mathematical quantity that can be represented by a single symbol.

Variance   A measure of the dispersion of a certain random variables.

Index

adaptation, 247–50
adjudication approach, 302–7
aerosols, 82–84, 182–83, 186–88
altitude, 141–42
ambient heat prejudice, 181
Anderson, David, 55–57, 235–37, 250
anomalies, 53, 125–26, 129, 170
Antarctica, 61–62, 259
A1FI file, 242–44
the Arctic
    average annual temperature series, 57–58
    greenhouse warming predictions, 53–54
    open water in, 40
assessment reports, 34, 43
atmosphere-ocean models (AOGCMs), 98
atmospheric carbon dioxide concentrations, 209–17
atmospheric temperature field, 136
attractor, 196
audits, 275–76
Australia, 51
autocorrelation coefficients, 202–3
average
    differences from, 53
    performance, 67–70
    types of, 107–8
averaged temperature "anomalies." *See* hockey stick graph
average temperature, computation of, 109–10, 119, 143–44. *See also* temperature
averaging, 65–66, 208

Babbitt, Bruce, 44
bell curve, 226
"beneficial costs" fallacy, 278–79
big initiatives, 28–29
Big Panel (United Nations)
    assessment reports, 34
    consensus, orchestration of, 46
    denunciation of critics, 31
    Little Panel, 237–38
    media treatment of draft summary, 36
    Official Science, representative of, 33
    Third Assessment Report, 34, 237, 242–44
    uncertainty of scientists, downplaying of, 43–44

    Working Group I, 29
Bleep, 127, 146, 148–49
Bohren, Craig, 301
Borehold-based global temperature statistic, 169
Brookhaven National Laboratory, 264
Brundtland Report on Sustainable Development, 37
Bulletin of the American Meteorological Society, 41–42
buoy network data, 131
Bush, George Jr., 48
butterfly effect, 74
Byrd Station, 215

Camp Century, 212
carbonated beverages, 269–70
"carbon cycle," 117
carbon dioxide
    and aerosols (*see* aerosols)
    atmospheric carbon dioxide concentrations, 209–17
    Byrd Station, 215
    Camp Century, 212
    damages estimate, 245
    estimate of future emissions, 240–41
    existing model effect of, 187–88
    Gigatonnes carbon equivalent (GtC), 210
    ice-core method, problems with, 213–16
    past levels of, 210–17
    plant growth and rising CO_2 levels, 264–65
    reconstructions of levels, 212
    Siple Station, 213–14
    statistical detection of causality, 223–30
    and surface warming, 217–23
    tax, 283–84, 285–86
    Taylor Dome, 212, 214–15
    tree leaves, 216–17
carbon sinks, 272–73
causation
    Gaussian density, 226–28, 230
    Granger causality, 224–25
    inferences about, 224
    Lévy flights, 228–29
    statistical detection of, 223–30
Center for the Study of Carbon Dioxide and Global Change, 264

central limit theorem, 227
certainty
  convection of. *See* convection of certainty
  Doctrine of Certainty. *See* Doctrine of
    Certainty
  self-reinforcing, 59
Challenger, 31
chaos, 66, 74, 199–200
chaotic radio waves, 175–76
chaotic system of equations, 195–96
Churchill, 58–59
cities, 129–30
climate
  and carbon dioxide, 187–88
  chaotic carrier, 197
  forecasts, 100
  "global" temperature, 100, 101–12
  importance of, 245–247
  observations, 77
  prediction, 113, 239–42
  and T-Rex ("global temperature"), 146–51
  vs. weather, 64
climate change
  "costs" of, 247
  and "forcing," 205–6
  remoteness of, 64
  temperature, focus on, 120
  temperature statistics and, 111
  U.S. project, 49–50
Climate Change Action Fund, 50, 148
climate forecasts, 46–47
climate models
  accuracy of, 201–5
  assumptions, 220
  atmosphere-ocean models (AOGCMs), 98
  effective emission altitude, 220, 221
  general circulation models (GCSs),
    95–101
  hierarchy of models, 92–95
  vs. meteorological models, 91–92
  natural variation, 203
  non-validation of, 99–100
  numeric computation, 15–16
  parameterizations, 219, 222
  regional circulation models (RCMs), 97
  6.5° C per km, 220–21, 222
  surface warming "predictions," 223
  usefulness of, 16–17
climate research, 41–42
Climate Research, 43
climate theory
  average performance, 67–70
  chaos, 66, 74

vs. climate models, 62
Enchanted Computing Machine, 76–80
Kolmogorov theory, 70–72, 75–76
Lorenz equations, 72–75
mathematics, 62–63
vs. metaphors, 62
Navier-Stokes equations, 66–67, 68–69, 71,
  72, 76
non-existence of, 63–76
questions about existence of, 61–63
turbulence, 69, 71, 72
closure problem of turbulence theory, 70
coal consumption, 241
computers
  and animation, 90
  fallibility, 88–89
  and mathematics, 88–89
convection in atmosphere, 221–22
convection of certainty
  described, 21
  the system, 39–53
"costs of inaction," 235–37

Daly, John, 262
data interpretation, 305
Dead Man's Isle benchmark, 262–63
degenerate signals, 205
degrees C. *See* hockey stick graph
density functions, 226–27
developing countries, 277–78
Devil's yarn ball, 80–87
differential equations, 73–74
dimensional analysis, 71
dissenters, 42–44
distributions, 226–27
Doctrine of Certainty
  averaged temperature "anomalies" (*see*
    hockey stick graph)
  and climate prediction, 113
  contrary to, 25–26
  damaging role of, 21
  described, 17–18, 19–20
  discussion of ideas in, 289–93
  and *Flying Dutchmen,* 308
  and greenhouse metaphor, 113–16
  and idiosyncratic language, 116–17
  irony, 25
  Official Science and governments, 39
  proposed measures, 268–70
  and science, 25
  scientific opposition to, 42
drunkard's walk, 228

effective emission altitude, 220, 221
effect without cause, 198–201
ELGs (environmental lobby groups)
    advice to politicians, 48
    and fundraising, 38
    selective information to public, 47
    and vertical movements of system, 44–45, 48
emission reduction policies, 283
empirical relationships, 91
Enchanted Computing Machine, 76–80
energy, 106
environmentalists
    industry subsidy of, 52
    as key players, 37–38
environmental lobby groups. *See* ELGs (environmental lobby groups)
Environment Canada, 118
equimarginal criterion, 282–83
ergodicity, 137
expertise, 29–30
extensive quantity, 108
Exxon, 52–53

fact-finding inquiries, 302
firsthand experience, 29–30
fluid dynamics
    and 1-D model, 94
    Navier-Stokes equations. (*see* Navier-Stokes equations)
    and rain, 84
    and thermodynamics, 82
    turbulent diffusion, 80–82
Flying Dutchmen, 308
fossil fuel firms, 38
fractal, 85–86
fractal geometry, 228
free-lunch fallacy, 279
function, 63

Gaussian density, 226–28, 230
general circulation models (GCSs), 95–101
geoengineering, 268
German consumer price index reconstruction, 172–73
glaciers, 150–51
Global Change Biology, 264
Global Historical Climatology Network, 138–40
Global Hydrology and Climate Center Web Site, 151
global temperature. *See* T-Rex ("global temperature")
global temperature statistic, 100, 101–12, 118

global warming
    best policy on, 293–97
    "consensus view," 31
    economic consequences of, 235
    effect of, 245–54
    vs. effect without cause, 198–201
    and heat, 181
    lack of proof, 193–98
    and rising sea levels, 257–64
    tempting cause, 28
global warming suspects
    aerosols, 186–88
    ambient heat prejudice, 181
    meteorological oscillations, 192–93
    the moon, 192
    solar activity, 190–91
    water vapour, 191–92
glossary, 312–14
Goddard Institute of Space Science, 130
good first step fallacy, 281–82
"grandmother test," 24
Granger, Clive, 224
Granger causality, 224–25
Gray, Vincent, 243
greenhouse effect
    "carbon cycle," 117
    entrenchment of, 20
    as media example, 24
    metaphor, 113–16
    and Pacific Ocean warming, 42
greenhouse gases, 117
Greenpeace, 47
grid boxes, 96–97
gridding, 127–28
gross domestic product reconstruction, 171–72
group polarization, 298–302, 303

Hadley Centre, 131
Hansen, James, 45
hidden worlds, 64–65
hierarchy of models, 92–95
Himalayas, 150
hockey stick graph
    artful joining, 157
    caption, 169
    described, 154–57
    extrapolation into past, 164–65
    fitting process, 160–61, 163–64
    maps and mappings, 159–64
    proxy data, 158–59
    reconstruction, 169–73
    singular value decomposition, 162
    and stationarity, 165–71

horizontal movements of system, 40–44
Houghton, Sir John, 33, 44, 45–46
How the News Makes us Dumb (Somerville), 36
humidex number, 246
humidity, 246

ice-core method, problems with, 213–16
the Idsos, 264
impacts research, 148, 234
infinite-sized computers, 78
insect-borne diseases, 250–51
intensive quantity, 108, 121
International Institute for Sustainable
   Development, 54
International Quaternary Research Society, 257
Inuit recollections, 54
irrelevant comparisons fallacy, 279

Jaworowski, Zbigniew, 214, 215
journalists. *See* media

key players
   environmentalists, 37–38
   industry, 38–39
   media, 34–37
   Official Science, 29–34
   public sector politicians, 26–29
kinetic energy rule, 110
kinetic theory, 67
knowledge, absence of, 209
knowledge gaps, 23–24
Kolmogorov theory, 70–72, 75–76
Kyoto Protocol
   Annex I nations, 271–72
   audits, 275–76
   Australian inquiry into, 51
   "beneficial costs" fallacy, 278–79
   Bush's rejection of, 48
   carbon dioxide tax, 283–84, 285–86
   carbon sinks, 272–73
   costs, 271, 278–82, 286–87
   "costs of inaction," 235–37
   emissions targets, 271–73
   Exxon's rejection of, 52–53
   fatal flaws, 287
   free-lunch fallacy, 279
   good first step fallacy, 281–82
   implementation benefits, 276–78
   irrelevant comparisons fallacy, 279
   leakages problems, 274–75
   meeting targets, costs of, 282–84
   Non-Annex I nations, 271–72
   permits, 272

precautionary principle fallacy, 280
total/marginal costs fallacy, 281
troubled life of, 270

latent heat, 80
law of group polarization, 298–302, 303
leakages problems, 274–75
Leipzig Declaration, 42–43
Lévy flights, 228–29
Lighthill, Sir James, 75
likeliness of future events, 254–57
Lindzen, Richard, 41–42, 45, 300–301
Little Panel, 237–38
logistic map, 199
Lorenz equations, 72–75, 88–89
Lucas, Robert, 218
Lucas Critique, 218

macroeconomic models, 217–19
Manning, Martin, 243–44
maps and mappings, 159–64
mathematics, 62–63, 79, 88–89, 309
media
   authoritarian aspect of, 35
   errors and corrections, 35
   global warming, treatment of, 36–37
   as key player, 34–37
   and motions of the system, 40
   newsworthy items, selection of, 41–42
   reporting standards for science, 24
   and science reporting, 35–36
Mendelsohn, Robert, 249
metaphors, 62, 113–18
meteorological models, 91–92
meteorological oscillations, 192–93, 198
methane emissions, 241–42
microwave radiation, 133
models. *See also* climate models
   artistry of, 90–91
   and computers, 87–88
   empirical relationships, 91
   vs. theories and metaphors, 62
   types, 89–90
the moon, 192
Mörner, Nils-Axel, 257–58
multicollinearity, 204–5

Nature journal, 31–32
Navier-Stokes equations, 66–67, 68–69, 71, 72,
   76, 86–87, 98
nescience
   carbon dioxide and surface warming,
   217–23

meaning of, 209
statistical detection of causality, 223–30
NGOs (non-government organizations). *See* ELGs (environmental lobby groups)
Nobel laureates, 12, 13
non-stationarity, 229
Nordhaus, Bill, 249
normal distribution, 226–27
North Atlantic Oscillation, 135

Oak Ridge National Laboratory, 264
oceans, 130–32
Official Science
A1FI scenario, 244
consensus, orchestration of, 46
described, 25–26
interaction with politicians, 39–40
as key player, 29–34
undermining science, 47
oil and gas industry
as key player, 38–39
subsidies by, 51–52
Orbital Engine Corporation, 51
ozone depletion, 37

Pacific Climate Shift, 135, 193, 199–200
Pacific Decadal Oscillation, 135
Pacific Ocean, arming of, 42
parameterizations, 16, 91, 97–98, 219
"peer review," 242–43
Petition Project, 43
phase locking, 18–19
polar bears, 54–55, 58
politicians. *See* public sector politicians
precautionary principle fallacy, 280
preferences for climate, 252–54
private sector politicians. *See* environmentalists
probability, computation of, 226–27
probability-density functions, 226–27
public sector politicians
advice from environmentalists, 48
interaction with Official Science, 39–40
key players, 26–29
and Official Science, 31
PUN. *See* Big Panel (United Nations)

radiative equilibrium, 222
radiative forcing of climate system
described, 179–80
measurement units, 180–83
"total" forcing effect, 185
radiative transfer, 84–85, 93–94
rain, 84–85

random walk, 228
reasons for book, 11–12
reconstruction, 169–73
redo averaging, 69
regional circulation models (RCMs), 97
Reynolds, Osborne, 69
Reynolds number, 71
Ricardian analysis, 248–49
Ricardo, David, 248
Ross-Lempriere benchmark, 262–63

Sachs Harbour, 54
Sandmo, Agnar, 285
satellites, 132–35
science
vs. civil authorities, 21–22
dissent, 42–44
and Doctrine of Certainty, 25, 42
firsthand experience, 29–30
journalists and, 24
by metaphor, 113–114
Official Science, 25–26, 29–34
opinions of scientists, 303–4
vs. politics, 26
scientist's burden, 21–26
undermining of, 47
Science, 32–33, 202–3
scientific journals, 31–32
Scripps Institute of Oceanography, 191, 192
sea levels
Antarctic ice, melting of, 259
background dynamics, 260–61
changes in, 257–61
Dead Man's Isle benchmark, 262–63
measurement of, 261–62
Shaw, Daigee, 249
Sierra Club, 47
signal detection, 174–76, 203–5
significance, 224–25, 226
simulacrum fallacy, 144–46
simulated scenarios, 239–41
Singer, S. Fred, 55
Siple Station, 213–14
6.5° C per km, 220–21, 222
size, 79
solar activity, 190–91
solar radiation, 268–69
Somerville, C. John, 36
soot emissions, 188
SPAMs, 33–34
spatial coverage, 136–38
Special Report on Emission Scenarios, 239
standard deviation, 225

stationarity, 165–71
statistical fitting, 160–61, 163–64
statistical modelling, 226–27
statistics, 102–3, 159–60, 197, 224–27
stepwise regression, 205
stomata, 216–17
sub-grid scale, 15
sub-grid scale phenomena, 97
subsiding infrastructure, 251–52
summaries for policy makers and media. *See*
SPAMs
Sunstein, Cass, 299, 302
superscripts in reports, 254–57
Suzuki, David, 47–48, 117
Svensmark, Henrik, 191
the system
and ELGs, 44–45
horizontal movements, 40–44
media, role of, 40
motions of, 39–53
and Official Science, 39–40, 45
and politicians, 39–40
vertical movements, 44–53
systems, 72

T-Rex ("global temperature")
autocorrelation coefficients, 202–3
Bleep, 127, 146, 148–49
cities, 129–30
climate, contrived connection with, 146–51
compromised continuity, 152
described, 122–23
hockey stick graph (*see* hockey stick graph)
inconsistency, 127–29
make-up of, 125–27
normal vs. anomaly, 125–26
oceans, 130–32
prediction of changes in, 239
sampling altitude, 141–42
sampling rule changes, 141
satellites, 132–35
as simulacrum fallacy, 145
spatial coverage, 136–38
station closure, 138–41
as statistic, 124
vs. temperature field, 245
and unnatural climate change, 177–78
Taylor Dome, 212, 214–15
temperature
average, problematic computation of,
108–10, 119, 143–44
and biology, 158–59
change, with altitude (6.5° C per km),
220–21, 222

vs. climate, 112
field, 105, 121, 245
global (*see* T-Rex ("global temperature"))
gridding, 127–28
intensive quantity, 108, 121
multiple temperatures, 121
no single value, 120, 143
normal vs. anomaly, 125–26
radiative equilibrium, 222
temporal detail, 137
theories, 62
thermodynamical theory, 80
Third Assessment Report, 34, 237, 242–44
thunderstorms, 14–15
total/marginal costs fallacy, 281
trade, role of, 249–50
tree leaves, 216–17
tree rings, 158–59
turbulence, 15, 69, 71, 72
turbulent diffusion, 80–82
turbulent flow, 229

UK Meteorological Office, 131
uncertainty
vs. absence of knowledge, 208
atmospheric carbon dioxide concentrations,
209–17
United Nations. *See* Big Panel (United Nations)
urban heat biases, 129
U.S. Standard Atmosphere, 220

variables, 63
vector autoregression, 224
vertical movements of system, 44–53
vortices, 71

Wagner, Friederike, 216
water vapour, 117, 191–92
Waterworld, 234
Watson, Robert, 33, 43–44
weather, vs. climate, 64
weather forecasts, 46–47
weather-related fatalities, decline in,
252–53
weather satellites, 132–35
West Antarctic Ice Sheet, 259
Wigley, Tom, 276–77
wind-chill factors, 246
World Bank, 245
World Wildlife Fund, 41

Yamal, 40

zero-space dimensions, 92